MARINE BIOLOGICAL INVASIONS
Ecology, Impacts and Management

PHILIPPE GOULLETQUER

Pelagic Publishing and Éditions Quæ

To cite this work:
Goulletquer P., 2025. *Marine biological invasions*,
Versailles, Éditions Quæ, 120 p.
https://doi.org/10.35690/978-2-7592-4150-7

This edition is a co-publication between
Éditions Quæ and Pelagic Publishing.

Pelagic Publishing
20–22 Wenlock Road
London N1 7GU
www.pelagicpublishing.com

Éditions Quæ conducts a scientific evaluation of manuscripts prior to publication,
the procedure for which is described here: https://www.quae.com/store/page/199/
processus-devaluation

The publication of this book has received financial support from Ifremer to enable it to
be distributed widely and openly.

The digital versions of this book are released under a CC-by-NC-ND 4.0 licence
(https://creativecommons.org/licenses/by-nc-nd/4.0/).

The figures or other third-party material in this publication are covered by the
publication's Creative Commons license, unless otherwise indicated in a reference to the
material. If the material is not covered by the publication's Creative Commons license
and the intended use is not permitted by law or exceeds the permitted use, permission
for use must be obtained directly from the copyright holder.

Paperback: 9781784276553
PDF: 9781784276560
ePub: 9781784276577

A CIP record for this book is available from the British Library

EU Authorised Representative: Easy Access System Europe –
Mustamae tee 50, 10621 Tallinn, Estonia, gpsr.requests@easproject.com

Editorial coordination: Aude Boufflet
Translation review: Helen McCombie,
Bureau de Traduction de l'Université de Bretagne Occidentale (BTU UBO)
Layout:  EliLoCom

Cover photo: Lionfish in the Red Sea, Egypt
© Reinhard Discherl/Biosphoto

Contents

Introduction—Are terrestrial invasive species getting all the media attention? 5

What do we know about biological invasions? 10
What do we mean by 'invasive species'? 10
How does a biological invasion take place? 14
Why should we care about invasive alien species? 17
What is the situation in mainland France and the French overseas territories? 23

How are exotic marine species introduced? 28
A historical perspective on species introductions 28
Current introduction vectors 32
Current routes of introduction 39

What are the impacts of biological invasions? 49
Characteristics and methods of impact assessment 50
Case studies 52
Importance of the ecosystem approach and analyses 63

How can we manage biological invasions? 65
Voluntary introductions 66
Unintentional or accidental introductions 69
Prevention and regulatory measures 70
Surveillance of entry routes 73
Raising public awareness of biological invasions 74
Once an invasive species is established what means can be used for its control? 75
How can marine invasive species be successfully eradicated? 83
Some cases of eradication 83

How can research and expertise contribute? 88
Assessing risks 88
Improving the detection of invasive alien species 91
Assessing the risks and impacts following an introduction 92
What are the expectations for research in the short and medium term? 97

Conclusion—Invasive marine species: what does the future hold? 101

References 106

INTRODUCTION
Are terrestrial invasive species getting all the media attention?

The introduction and spread of alien (or 'exotic') species, leading to biological invasions, is one of the major causes of biodiversity loss worldwide. These invasive alien species (IAS) can, in some cases, compete with local species, modify environmental conditions and the services provided by the environment (ecosystem services), or damage economic activities and human health. A study published in 2021 in the scientific journal *Global Change Ecology* showed that 14% and 40% of functional diversity (habitats and biomass) for mammals and birds, respectively, were threatened by biological invasions (Bellard *et al.*, 2021). For the European Union (EU) alone, the economic costs of such impacts are estimated at more than 138 billion euros (€bn) for the period 1960–2020, of which nearly €8bn is fully documented, with 10.28% allocated to management costs. Projections for 2040 suggest a further deterioration, with costs expected to reach at least €22 billion for the EU, including €2.4 billion for France (Henry *et al.*, 2023). This issue has therefore become a major concern for many land managers and for the development of public policies.

The issue is covered by all the international conventions dealing with environmental and development issues. The phenomenon is particularly serious in island systems, where the introduction of cats and rats, for example, is responsible for the extinction of endemic species. The disappearance of the dodo (*Raphus cucullatus*), an endemic bird of Mauritius, at the end of the 17th century is a case in point. Its extinction is the direct result of human activities, as it fell victim to hunting activities, changes to the soil caused by imported crops and the predation of eggs by various exotic species that were also imported (rats, dogs, cattle). Today, the kagu (*Rhynochetos jubatus*), a phylogenetically unique bird species emblematic of New Caledonia, is similarly threatened with extinction by rat predation. Moreover, these changes can also be documented on a continental scale. For example, the case of the multiple introductions of exotic earthworms into North America, whose initial biodiversity had declined sharply following the last ice age, is also exemplary. More than 97% of the North American continent has been colonised by 70 species of earthworm originating from Asia and Europe, representing a quarter of this biodiversity, which plays a major ecological role (Mathieu *et al.*, 2024).

For some ecosystems that are major contributors of food resources, voluntary introductions have proved catastrophic. In Africa, for example, the Nile perch (*Lates niloticus*), a powerful carnivore and very good swimmer, was introduced into Lake Victoria in 1954 to counter the collapse of the population of Victoria tilapia (*Oreochromis variabilis*) and Singida tilapia (*O. esculentus*), two species naturally present in the lake. Initially seen as the 'saviour' fish, contributing to the diet of more than 47 million people in the three neighbouring countries (Uganda, Tanzania and Kenya), this introduction has proved problematic for the environment, with the disappearance of more than 200 species of native fish and the disruption of the entire ecosystem. After peaking at almost 380,000 tonnes in 1990, landings have fallen to just under 200,000 tonnes since 2020. It should be noted that production is supplemented by the exploitation of another species introduced at the same time, the Nile tilapia (*O. niloticus*).

France is directly concerned by the problem, with numerous examples in both mainland France and the overseas territories, with more than 3,700 exotic species to date in France overall (there are 1,459 species on average per EU country) (Henry *et al.*, 2023). The French Biodiversity Agency (*Observatoire National de la Biodiversité*: ONB) indicator shows that an average of 12 IAS are introduced into a French *département* every decade, and that the rate is increasing. This threat is particularly acute in French overseas territories, where 74% of French IAS are concentrated, causing irreversible damage to local and endemic flora and fauna. Furthermore, 60% of the world's 100 most invasive species have been identified in these overseas territories.

These include the *Miconia calvescens* tree, nicknamed the 'green cancer' of Tahiti, which is proliferating at great speed, to the detriment of the local flora. Introduced in 1937 as an ornamental plant, it now covers two-thirds of the island (Meyer, 2023).

Nowadays, the issue of biological invasions is appearing more and more in public debate and in major national and regional media, especially in connection with a few emblematic cases. The coypu (*Myocastor coypus*), American mink (*Neovison vison*) red-eared terrapin (*Trachemys scripta elegans*), common wall gecko (*Tarentola mauritanica*) — which arrived in the south of France in the early 1980s and has gradually colonised the entire coastline, replacing the common wall lizard (*Podarcis muralis*) — American bullfrog (*Lithobates catesbeianus*), ring-necked parakeet (*Psittacula krameri*), Japanese knotweed (*Reynoutria japonica*), water primrose (*Ludwigia* sp.), mimosa (*Acacia dealbata*), frogbit (*Limnobium laevigatum*), pampas grass (*Cortaderia selloana*), native to South America, pickeral weed (*Pontederia cordata*) and water hyacinth (*Eichhornia crassipes*) are all invasive species that threaten the environment, the economy and, in some cases, public health. Some threaten human health by carrying diseases, such as the tiger mosquito, which carries the dengue and chikungunya viruses, or by causing allergies, such as ragweed, which causes conjunctivitis, asthma and urticaria, or by being toxic to humans, such as giant hogweed (which causes burns). This problem affects the vast majority of plant and animal groups. Exotic plants such as the sour fig (*Carpobrotus edulis*), prickly pear (*Opuntia ficus-indica*) and century plant (*Agave americana*) have proliferated to such an extent that 200 tonnes of them had to be removed (at great expense) from the calanques of Marseille between 2017 and 2022 in order to protect astragalus (*Astragalus tragacantha*), an emblematic plant of the area. The number of projects to dig out exotic plants, involving the general public, is increasing in France. Sour fig, Himalayan balsam (*Impatiens glandulifera*) and Japanese knotweed were dug out in the Lannion-Trégor community (Côtes-d'Armor) in 2024 to preserve coastal biodiversity.

The most recent cases, which have also received a great deal of media coverage, include the Asian hornet (*Vespa velutina*), which was recently joined by the Oriental hornet (*Vespa orientalis*) in 2021 (Marseille, 2021), the virile crayfish (*Fraxionus virilis*) (Yonne, 2021), electric ant (*Wasmannia auropunctata*) (Toulon, 2022), and tiger mosquito (*Aedes albopictus*). American crayfish, particularly the Louisiana crayfish (*Procambarus clarkii*), are responsible for a wide range of damage, both to biodiversity (direct competition with many animal species) and to habitats, due in particular to their burrowing activity. In 2024, following heavy early-summer rains, the red swamp crayfish, a species introduced in the late 1970s for commercial purposes, invaded several areas of the Atlantic coast, spilling out onto roads, gardens, car parks, etc. Population densities were so high that the crayfish saturated available space and pushed into new territories. Hardy and voracious, this species, having already displaced other non-native crayfish, continues to expand exponentially through the country's freshwater systems, disrupting ecosystems by preying on amphibian eggs and young fish, and by digging burrows that erode riverbanks.

Although the problems are similar, biological invasions in the marine environment do not attract as much attention as those on land. In less accessible environments, ecosystem characteristics are less well studied, which makes marine IAS, primarily coastal macrofauna and macroflora larger than 1 mm, less visible. However, a few cases have been highlighted in order to inform the public about the threats that primarily affect either public health or human activities, especially fishing. The brown seaweed *Rugulopteryx okamurae*, which arrived in the 2000s and originated in Japan, is now present throughout the Parc des Calanques and in several Mediterranean areas. Here, it is transforming the habitat by completely covering the rocks, thereby creating a significant change in the marine flora and fauna (Ruitton *et al.*, 2021). This species also arrived in Gibraltar in 2015 and has colonised Spanish waters as far as the Canaries and the Basque Country, causing considerable damage to biodiversity and the fishing industry. As with 'green tides', its degradation forms a 'bank' of biomass and, as it rots, releases hydrogen sulphide, which is harmful to humans (García-Gómez *et al.*, 2021a).

The American blue crab (*Callinectes sapidus*) is also a species of great concern in the Mediterranean, both for the environment and for human activities. Some species of fish are also reported in the media because of their significant impact: the square-tailed rabbitfish (*Siganus luridus*) owes its name not to its appearance, but to its diet. It is a highly efficient herbivorous fish native to the Indian Ocean, capable of ravaging the seabed and profoundly altering its environment. The invasion of Caribbean waters by red or common lionfish (*Pterois volitans* or *P. miles*) has disrupted the entire ecosystem of the region, especially the structure of coral reefs. Similar impacts have been observed in the eastern Mediterranean since the arrival of *P. miles*. This species is of particular interest because of its expansion towards the western Mediterranean. For more insidious and invisible reasons, the proliferation of ostreopsis (*Ostreopsis ovata*), a microalga of tropical origin, is also attracting attention because it is responsible for toxins dispersed as aerosols via sea spray. These toxins can contaminate beach users by inhalation, causing symptoms that are often similar to flu. This microalga has already caused beach closures and hospitalisations in the Mediterranean and on the Basque coast in recent years. Alongside these few high-profile species, several hundred exotic species have been present on coasts of mainland France and French overseas territories for many decades, or even centuries, some having become part of our natural heritage, others being exploited. However, many have arrived on our coasts more recently (Goulletquer, 2016). Every year, new reports that could lead to new biological invasions are recorded, such as the red alga *Lophocladia lallemandii*, identified in the Port-Cros national park in 2021. Several dozen species are either directly responsible for, or commonly associated with, significant impacts on the environment and/or human activities.

The aim of this book is to throw light on the process of marine biological introductions and invasions, specifying the vectors and pathways of introduction, the impacts caused, and the management methods implemented to meet this challenge. The contribution of all aspects of scientific research is essential here, in order to provide the most convincing results for managers and public decision-makers, who will be able to draw up new public policies and regulations at national, European and international level.

WHAT DO WE KNOW ABOUT BIOLOGICAL INVASIONS?

WHAT DO WE MEAN BY 'INVASIVE SPECIES'?

Although the terminology is not yet fully established, it is important to clarify the terms used in the field of biological invasions. (Soto *et al.*, 2024; Vilizzi *et al.*, 2025). This is a complex subject with a variety of semantics, due in particular to the different cultural perceptions of 'man-nature' relationships.

For example, there are various terms for introduced species, such as 'alien', 'exotic', 'non-native', 'non-indigenous', 'allochthonous' and 'xenobiotic', often used synonymously and sometimes depending on the context. They refer to species that have been intentionally or accidentally transported by human activity to a region where they were not originally naturally present (i.e. outside the species' historical native range). This implies a break in the species' natural range, for example a species naturally present and initially described in the Caribbean, its native range, being found and identified in the Mediterranean Sea.

Conversely, a change in the range of a species resulting, for example, from climate change, which is an increasingly frequent situation, does not correspond to an introduction, since there is no break in the distribution range. These are referred to as neo-native species (Essl *et al.*, 2019).

Similarly, it is necessary to avoid any parallels with issues of human immigration, as sometimes discussed in the social sciences, insofar as human population movements do not meet the criterion of a break in distribution area. *Homo sapiens* has been present for several millennia throughout the world, with the exception of there never having been perennial human populations in Antarctica (Rémy and Beck, 2008; Warren, 2021)!

Some species are described taxonomically, but their natural area of origin cannot be determined. These are known as cryptogenic species, whose origin is unknown (Carlton, 1996; Jaric *et al.*, 2019). A typical case would be the inventorying of previously undescribed species present in the biofouling on the hull of a merchant ship that had transited several continents before being refitted in a dry dock in a European port.

In a more complex way, certain groups of species that are not morphologically distinct may meet the definition of 'species' through reproductive isolation, or the phylogenetic definition of species (strong genetic differentiation of lineages due to ancient divergence). A complex of native and exotic species, known as 'cryptics', can thus be found, which justifies the widespread use of genomic approaches to characterisation beyond morphological criteria alone (Jaric *et al.*, 2019).

Some of these introduced (non-native/exotic/non-indigenous) species survive and establish natural populations, and a fraction of them may become invasive. The term 'invasive' applies to exotic/non-native species that spread, with or without human assistance, in natural or semi-natural habitats. These species induce a significant change in the composition, structure and functionality of ecosystems and/or cause significant economic losses and/or have effects on human well-being and public health and, ultimately, induce additional management costs. The term 'invasive' has a strong connotation of urgency, risk and negative impact. Some definitions, such as that of the United Nations Environment Programme (UNEP) in 1994, restrict the 'invasive' characteristic to species that have a negative impact on host ecosystems. However, this criterion of negative impact can be subjective and relative, because it is anthropocentric. Even if the vast majority of impacts are considered 'negative', it is necessary to consider all the effects on biodiversity and on the ecosystem services produced, including both 'negative' and 'positive' (e.g. supply services) (Kourantidou *et al.*, 2022; Tsirintanis *et al.*, 2022). For example, the proliferation of the Manila clam (*Ruditapes philippinarum*) has had a 'positive' effect on the winter survival rates of marine avifauna on British coasts, through improved availability of prey (Caldow *et al.*, 2007).

The creation of habitats by so-called 'engineer' species can be 'positive' in certain respects, such as the increased availability of refuges for other local species, but 'negative' in others, considering the impact on the original natural habitat, or even by favouring the arrival of new exotic species (e.g. novel ecosystem) (Tsirintanis *et al.*, 2022).

As far as marine biological invasions are concerned, we will stay with the category of invasive non-native alien species (IAS) insofar as the management methods differ profoundly between native and non-native species. Green tides, for example, which have had a great deal of media coverage, are environmental problems linked to the eutrophication of environments. They call for upstream management measures at catchment basin level, whereas IAS are a matter of their own characteristics as they develop in a new environment. From the point of view of management methods and regulations, the issue is also very different. Directive 2008/56/EC of the European Parliament and of the Council of 17 June 2008, known as the Marine Strategy Framework Directive (MSFD), identifies two distinct descriptors, No. 2 for 'non-indigenous species' (NIS) and No. 4 for 'eutrophication' in the case of green tides. Similarly, EU regulation 1143/2014 is dedicated solely to the "prevention and management of the introduction and spread of invasive alien species".[1]

A number of other terms are used to describe the different types of non-native species. Some non-native species may be observed from time to time ('occasional species'). These reports refer to taxa (species, subspecies, race, variety) introduced without the development of a perennial population. For example, the breeding of kuruma shrimp (*Penaeus japonicus*) during the summer in the maritime marshes of the French Atlantic coast has led to escapes and to reports of individuals in open environments, but no wild populations have appeared to date. This is also the case for the American blue crab on the same Atlantic coast, which has been occasionally observed following deballasting

---

1. https://eur-lex.europa.eu/legal-content/FR/TXT/PDF/?uri=CELEX:32014R1143

since the beginning of the 20th century (Goulletquer, 2016). What remains to be done is to analyse their future in the face of changes resulting from climate change!

Populations of 'established/acclimatised' taxa refer to the processes followed by a non-native species developing perennial populations following its introduction and successful reproduction. This is the initial stage that precedes 'naturalisation', when such a species becomes permanently established in its environment and integrated into the local ecosystem after several generations. Consequently, a naturalised species will be successful when it has overcome the following three barriers: geographical displacement, resistance to local environmental barriers and regular reproduction over time.

'Feral' populations refer to organisms, or their descendants, derived from escapes and having developed perennial populations after reproduction. Several non-native species used in aquaculture during the 20th century have developed such wild populations on the Atlantic coast: the hard clam (*Mercenaria mercenaria*), Manila clam, and Pacific oyster (*Magallana gigas*), formerly known as *Crassostrea gigas* (Goulletquer and Héral, 1997). Nowadays, these species are caught both professionally and recreationally. Wakame (*Undaria pinnatifida*), a macroalga native to Asia, was detected in Thau lagoon in 1971 and associated with the introduction of *M. gigas*. It was deliberately introduced into Brittany in 1983 for seaweed farming. After escaping from its cultivation area, this macroalga developed perennial populations on Brittany's coasts (Voisin *et al.*, 2007). Today, it is present on the coasts of Ireland, Scotland, the Netherlands and as far south as Spain as a result of seaweed farming activities and secondary introductions (Epstein and Smale, 2017). Escaped farmed individuals may also hybridise with wild populations, thereby altering their genetic characteristics, as observed in Atlantic salmon (*Salmo salar*) (Perriman *et al.*, 2022). Currently, these difficulties have led to a call for a halt to salmon farming in Canada in response to the 'endangered' status of wild, native populations of chum salmon (*Oncorhynchus keta*).

Another situation that needs to be clarified is 'translocation', which refers to the introduction of a species native to a geographical area within a country into another area of the same country where it is not native. The different coastlines of mainland France provide examples of this type. For instance, the voluntary translocation of Mediterranean mussel spat (*Mytilus galloprovincialis*) to the Normandy coast for farming there was carried out in the 1990s. The case of the accidental introduction of *Tritia neritea*, a nassariid gastropod, is also a good example. A study of the genetic structure of this mollusc, whose native range extends from the Mediterranean to the Atlantic coast of Morocco and southern Spain, has shown translocations to the French Atlantic coast and the English Channel via movements of shellfish stocks since the 1970s. Since then, it has been competing with the local species, the netted dog whelk (*Nassarius reticulatus*) (Simon-Bouhet *et al.*, 2016; Boissin *et al.*, 2020). This situation is important in terms of management when we consider the official reference lists of IAS, but also of protected species. These lists are drawn up on a national scale by means of single lists, with no distinction being made between occurrences on the different coastlines. The issue is even more complex when a species is the subject of protection measures or has 'endangered' status in its native range.

HOW DOES A BIOLOGICAL INVASION TAKE PLACE?

Biological invasion should be seen as a process that enables a species to break through 'barriers'. The introduction of individuals or reproductive elements (eggs, propagules) enables an initial geographical barrier to be crossed via direct or indirect human-mediated vectors, going beyond the species' natural range (e.g. maritime transport, Suez Canal). Subsequent dispersal, known as 'secondary' dispersal, may be facilitated by mechanisms and circumstances such as changes in the physical habitat, hydrological regime, physico-chemical characteristics and connectivity, as well as induced effects on populations and genetic and ecosystem impacts.

Figure 1 sets out the various stages in the process leading to a biological invasion and the potential management options. The different research approaches and actions are also outlined.

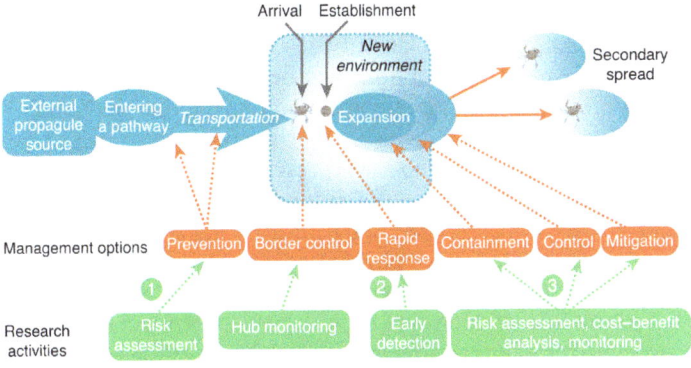

Figure 1. Typical diagram of a biological invasion process (in blue) and the different ways in which it can be managed (in orange), as well as research activities in response (in green).

Source: based on Olenin *et al.*, 2011, © 2011 Elsevier Ltd. All rights reserved, blue crab illustration: Tracey Saxby, Integration and Application Network (ian.umces.edu/media-library) (https://creativecommons.org/licenses/by-sa/4.0/), with permission from Elsevier.

A theoretical but realistic example illustrates the point: a merchant ship takes on its cargo in the port of Baltimore (USA) and stabilises its buoyancy by ballasting with seawater from the port. In fact, it simultaneously takes on many species that are present locally. It then crosses the Atlantic and arrives in Le Havre, France, where it unloads its cargo and changes the water in its ballast tanks (Arrival). The species are released in the port, where only a fraction survive the new environmental conditions. However, in the absence of natural controlling factors in this new environment, like predation, parasitism or even disease, a small fraction will not only survive, but will reproduce, developing a local population (Settlement) and potentially become invasive (Propagation). Another vessel will later contribute to the dispersal of these exotic species to another destination by the same processes (Secondary dispersal), where they will weaken the ecosystem and local biodiversity as a result of their proliferation. In terms of management, ballast water treatment on board can limit the initial introduction (Prevention).

Failing that, an operational surveillance network (Monitoring) organised at the level of the port can rapidly detect new species known to be invasive and initiate an action plan (Rapid Response) as long as the IAS population remains limited in number and surface area. Beyond that, more extensive containment measures (e.g. closing locks to isolate a basin) are still possible. Failing that, the management options will be reduced and will focus solely on limiting the development of populations (reduction, annual destruction management plan). For research activities, the priorities are to understand the processes involved in controlling the vectors of introduction, for example defining protocols and standards for treating ballast water and assessing the risks in order to prioritise the species to be targeted. Operational surveillance strategies (e.g. protocols, identification, sampling effort, prioritisation of introduction sites/points, new eDNA technologies) and rapid response (treatment protocols/eradication strategies) are all scientific elements that can be made available to managers to facilitate decision-making and contribute to the development of public policies (Olenin *et al.*, 2009; 2011).

Different situations may arise, depending on the species and environment concerned. In particular, the time required for each phase can vary considerably (**Figure 2**). For example, the latency phase once the species has been introduced can be very short or last several years, or even decades. The case of the Japanese oyster drill (*Ocinebrellus inornatus*), a predator of farmed shellfish, is of interest: genetic analyses have linked the presence of this Asian species to the mass introduction of Pacific oysters in the early 1970s, although it was only identified in 1994 on Ile de Ré (Pigeot *et al.*, 2000; Martel *et al.*, 2004a). The warming of the marine environment at that time facilitated its demographic explosion, with consequent impacts on the mortality of oysters in oyster beds. The expansion phase along the Atlantic coast was facilitated by oyster farming transfers between production basins (Martel *et al.*, 2004b). In addition, some species such as the common slipper limpet (*Crepidula fornicata*) are still invasive a century after their introduction, although some geographically localised populations have declined. Conversely, other species have become 'integrated' into the natural biodiversity or are declining/disappearing after a phase of massive proliferation. This is the case of *Caulerpa taxifolia* in the Mediterranean Sea, a seaweed that declined sharply since its accidental introduction in 1984, followed by a spectacular invasion lasting until 2007, when an as yet unexplained sharp decline was observed. It should be noted that two other species of *Caulerpa*, *C. taxifolia* var. *distichophylla* and *C. cylindracea*, both of Australian origin and both invasive in nature, have also appeared in the Mediterranean (Piazzi *et al.*, 2005; Picciotto *et al.*, 2016). Between these two extremes, intermediate scenarios may arise, making it more difficult for managers to make decisions and implement management schemes.

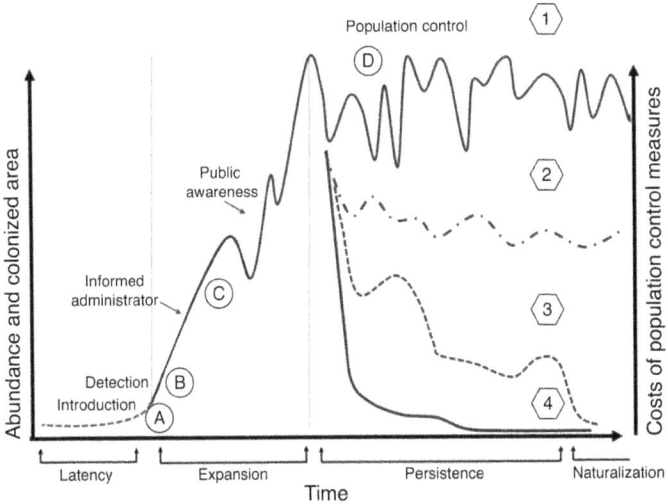

Figure 2. Temporal dynamics of a biological invasion: different types of biological invasion and potential management options. (A) Simple eradication, (B) eradication still possible, (C) eradication difficult to impossible, (D) management options only. (1 to 4) Different scenarios: from invasiveness persisting over time (1) to (4), characterised by rapid collapse followed by integration into the natural biodiversity.

Source: © 2025, Philippe Goulletquer.

WHY SHOULD WE CARE ABOUT INVASIVE ALIEN SPECIES?

The United Nations Conference on Environment and Development, better known as the Rio Summit, adopted a declaration in 1992 setting out the rights and responsibilities of countries in the field of the environment. Since then, international initiatives have multiplied to encourage countries to commit to finding solutions to the major environmental issues affecting the world. Three 'sister' conventions, all intrinsically linked, were adopted in the wake of the Summit: the United Nations Framework Convention on Climate Change (UNFCCC), Convention on Biological Diversity (CBD) and United Nations Convention to Combat Desertification (UNCCD). Coordination has been strengthened by establishing a liaison group that also includes the Ramsar Convention on Wetlands (1971), to develop synergies in their activities on issues of mutual interest. Panels of scientific experts report on the latest advances and knowledge in the field(s) concerned. This scientific liaison is essential for alerting decision-makers and civil society and is indispensable for international negotiations. This is the case of the Intergovernmental Panel on Climate Change (IPCC) for climate, and the Intergovernmental Science-Policy Platform on Biodiversity and Ecosystem Services (IPBES) for biodiversity. The two ongoing crises, the climate emergency and the erosion of biodiversity, explain the need for an energy transition and transformative changes for biodiversity. Specific initiatives — on a global scale, such as the Millennium Ecosystem Assessment (MEA, 2005) — have aimed to measure the impact of the transformations undergone by the ecosystems on which our survival and well-being depend. Over a period of five years, the MEA brought together more than 1,500 experts on a regular basis to identify the main pressures, propose a conceptual framework and formulate forward-looking scenarios to feed into decision-making and the development of national action plans (**Figure 3**). This conceptual framework was subsequently converted into operational terms at European level by the MAES ecosystem services mapping assessment project (MAES, 2020), and then into the IPBES conceptual framework at global level (**Figure 4**). The MAES project formed the foundation for the French programme *Évaluation française des écosystèmes et des services écosystémiques* (Efese), which produced an assessment of marine and coastal environments in 2018.

Over and above the fact that biodiversity is the foundation of the goods and services provided by nature, the MEA has identified the major causes of environmental disruption as the disappearance and degradation of natural environments, the over-exploitation of natural resources, the introduction of exotic species and global change, including climate change.

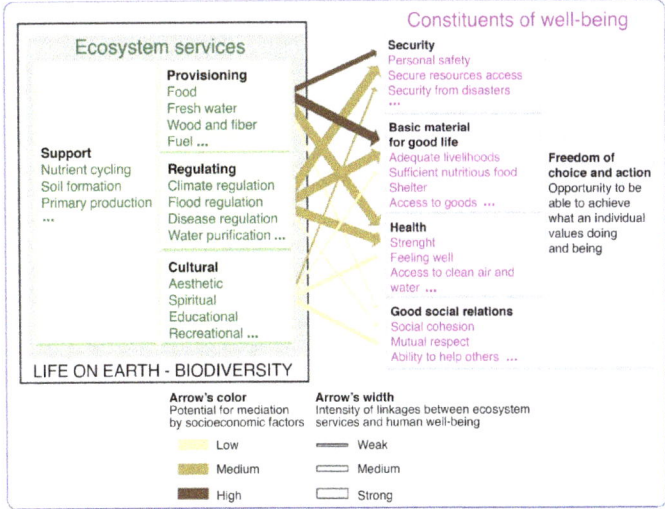

Figure 3. MEA conceptual framework specifying the services provided by nature that interact with human well-being (intensity of interactions and modulation/management capacity).

Source: based on Millennium Ecosytem Assessment, 2005 © Millenium Ecosystem Assessment.

All these pressures also apply to the marine environment. Marine ecosystems are mainly undermined by human activities through the introduction of exotic species, which are responsible for biological invasions, the over-exploitation of resources (e.g. fisheries), pollution issues such as plastics and associated molecules, eutrophication resulting from the enrichment of the environment (e.g. green tides), the alteration of habitats (e.g. coastal erosion) and climate change, which acts in its own right but also in synergy with these other factors. The loss of wetlands and the decline of mangroves, coastal erosion, the depletion of coral impacted by heat waves and acidification, pollution (plastics), the problems of overfishing and illegal and undeclared fishing, changes in the distribution areas of species as a result of climate

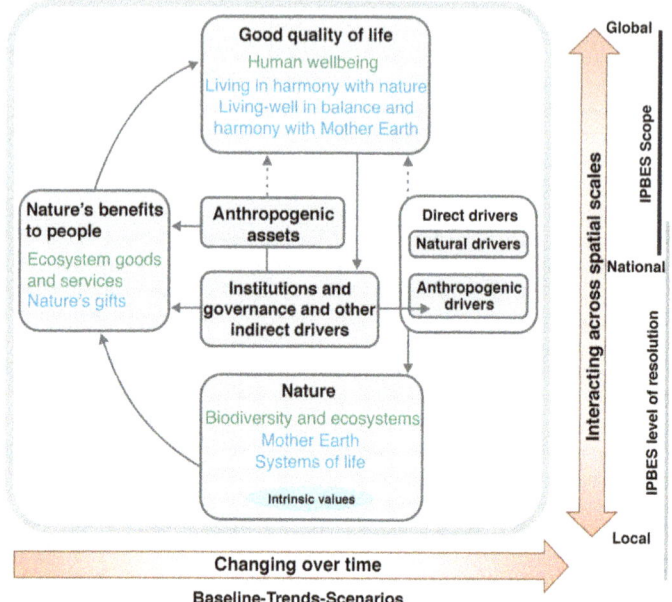

Figure 4. IPBES conceptual framework.

Source: Diaz et al., 2015, © 2014 The Authors. Published by Elsevier Ltd (http://creativecommons.org/licenses/by-nc-sa/3.0/), with permission from Elsevier.

change, and the tropicalisation of the Mediterranean Sea are all indicators of these pressures.

The nature of these pressures is also confirmed by the results of the CBD, which identifies biological invasions as the second most important cause of biodiversity loss, and by the global assessment carried out by the IPBES (2019). In fact, all international, regional (EU) and national strategies aim to help biodiversity recovery by reducing the pressures on it. These issues are reflected in European and national strategies and in EU directives and regulations (e.g. MSFD, 2014 regulation) as well as in the recent European regulation on nature restoration,[2] which represents a commitment to the application of the

---

2. https://eur-lex.europa.eu/legal-content/FR/TXT/HTML/?uri=CELEX:52022PC0304

Global Biodiversity Framework adopted at the Conference of the Parties (COP 15) in Montreal (CBD, 2022). Under this agreement, the international community set itself the ambitious target of reducing the rate of introduction of IAS by 50% by 2030. These guidelines are being transformed into strategies and action plans at the national level.

The issue of biological invasions is particularly important here as a direct factor in the loss of biodiversity on a global scale, as identified by the CBD in 2002. This is particularly true for oceanic islands, where IAS are considered to be the primary cause of species extinctions and the transformation of ecosystems. Several of them can be harmful to human health and, by definition, all have either economic or environmental impacts.

As a result, species introductions are one of the major ecological problems of our century. This phenomenon is all the more worrying in that, given our current state of technology and knowledge, they are almost always irreversible in the marine environment. Given the increase in commercial and human flows on an international scale, this problem is likely to get worse in the future. Furthermore, it is a largely invisible yet very real threat, often poorly understood and insufficiently examined.

The public does not necessarily have an accurate perception of the history of the presence of exotic species in their natural environment: some exotic species may be perceived today as an integral part of the natural heritage of our coasts. What fisherman of shellfish such as the soft-shell clam (*Mya arenaria*) or Pacific oyster on the Breton coast suspects that these species are not native here, having seen them there since his early childhood? This is known in fisheries management as the 'shifting baseline' syndrome, whereby the state of stocks and their specific composition are assessed on a current basis, without taking into account the extent of historical changes, and in fact underestimating them. The same applies to exotic species.

Although the introduction of exotic species by human activity is a phenomenon that has been going on for hundreds of years, if not millennia, this process has been growing in intensity for several decades. Annual rates of introduction and the number of established species have increased for most taxonomic groups and on all continents, particularly over the last fifty years. Awareness that biological invasions are an essential component of global change has recently increased worldwide. As a result, this pressure has only recently come to be seen as a determining factor in the changes undergone by the marine environment and in the services it provides.

Few official inventories exist on a global scale. Coordinated by the International Union for Conservation of Nature (IUCN), the Global Invasive Species Database (GISD) documents more than 1,000 marine invasive species in detail. Recently, the IPBES identified more than 37,000 alien species introduced to places all over the world, across all ecosystems (terrestrial, aquatic and marine), of which 3,500 are particularly invasive and responsible for more than 60% of recent extinctions of local species populations (IPBES, 2023). This latest analysis, prepared by 86 international experts in the field from 49 countries, was based on a census of 13,000 reference studies synthesised over four years. The economic consequences of IAS were estimated in 2021 at a cumulative value of at least $1,288 billion since 1970. Costs have quadrupled every decade since 1970. Spending on preventing or controlling these biological invasions represents only 1% to 10% of these costs (Leroy *et al.*, 2022, InvaCost project).[3]

In European ecosystems, more than 14,000 exotic species have been identified across all environments, 1,000 of which are considered to have a high impact due to their invasive nature. Given this number, management priorities are necessary. Projections are not encouraging either, with Europe expected to be the most affected by 2050 due to the effects of climate change (Biodiversa, 2017).

The harmful effects and nuisances caused by these species may only emerge long after their initial introduction. However, different scenarios can be observed: a regression or even a natural and gradual disappearance of the IAS, its naturalisation by integration into the local ecosystem or, conversely, the persistence of its invasive nature even after several decades. These species

---

3. https://invacost.fr/

often become established in ecosystems over the long term, changing the local environment by modifying the structure and functioning of the ecosystem.

This need to give greater consideration to the problem of marine biological invasions must be viewed in the current international context, shaped by the United Nations Sustainable Development Goals, particularly Goal 14 on the marine environment, and within the framework of the Decade of the Ocean, with the 2024 Barcelona Declaration setting future priorities for knowledge production and ocean science. This includes the co-design and co-delivery of science and knowledge, initially to understand global distribution, human health and the impacts of marine pollution on ecosystems, then to strengthen sustainable aquatic food production and finally to encourage sustainable ocean economy projects that are resilient to climate change. International initiatives are currently underway, including the Partnership for Observation of the Global Ocean (POGO), whose Ocean Biomolecular Observing Network (OBON) should help to improve knowledge of exotic and invasive species.

WHAT IS THE SITUATION IN MAINLAND FRANCE AND THE FRENCH OVERSEAS TERRITORIES?

Unlike a number of countries, France does not have an operational network, like that shown in **Figure 1**, that would enable rapid detection of exotic species and hence the possibility of corrective action plans. The current approach is based mainly on the observations by scientists carrying out research projects and/or fauna and flora inventories in certain geographical areas. Specific events, such as the massive deliberate introduction of the Pacific oyster in the 1970s, have been the subject of detailed analyses of the associated exotic fauna and flora (Gruet *et al.*, 1976). International working groups such as ICES-ITMO (International Council for the Exploration of the Sea-Introductions and Transfers of Marine Organisms) annually list the cases of introduction in the North Atlantic by country. Observations have been only progressively compiled to produce an inventory of the French coastline. As a result, the number of exotic species identified is underestimated. The situation is tending to improve with the implementation of international regulations, in particular with the MSFD, which aims to re-establish a 'good ecological status' of European waters, and for which one of the descriptors (No. 2) is directly dedicated to 'non-indigenous species'. A national initiative has been launched in recent years to improve scientific coordination (Massé *et al.*, 2023). The 2014 EU regulation reinforces the need for monitoring and risk analysis of exotic species for the purposes of reporting at the European level and improving the marine environment. As a result, 'high-risk' sites are beginning to be specifically monitored.

In the absence of a strict monitoring system for the marine environment, it is important to highlight the efforts made by associations in the field of participatory science to identify and inventory exotic species, mainly macrofauna and macroflora, as well as the initiatives of regional authorities. A non-exclusive list includes *Observation de l'Environnement en Bretagne* (OEB), the *Office de l'Environnement de la Corse* (OEC), dedicated observation networks in Normandy including port monitoring, and networks in Corsica and Occitanie (*Sentinelles de la mer Occitanie*). National coordination of six regional 'alien networks' was under construction in 2024 (OFB, 2024). The *Inventaire National du Patrimoine Naturel* (INPN[4]) is the reference platform on the status and conservation of French biodiversity and geodiversity (France and overseas territories), for all ecosystems combined. The site includes information on IAS and the observations of the six 'alien networks'.

Although the majority of the country's biodiversity is found in the overseas territories, there is a huge lack of knowledge about marine IAS, and inventories are very patchy. The work carried out by environmental associations, research bodies and local authorities is important. An initial assessment of marine biological invasions in French overseas territories has been carried out by a working group coordinated by the French Committee of

---

4. https://inpn.mnhn.fr (site under maintenance in November 2025)

the IUCN (2019). The analysis highlights the growing threat posed by IAS and, to date, identifies 59 exotic species and 32 cryptogenic species across the 13 French administrative regions, with ascidians being the most numerous (22). Amongst these species, lionfish and the seagrass *Halophila stipulacea* are invasive in the French West Indies, as is the green alga *Ulva ohnoi* in New Caledonia, the crown of thorns starfish *Acanthaster planci* in French Polynesia and New Caledonia, and the European green shore crab *Carcinus maenas*, which is on the list of the 100 most invasive species in the world, in Saint Pierre and Miquelon. The report also identifies the exotic species of risk in the countries bordering these administrative regions, where there is regular trade, making it possible to anticipate future arrivals.

The most recent analysis at the scale of mainland France resulted from the coordination of a group of researchers involved in this field (Massé *et al.*, 2023). This study updated the inventory of exotic marine species in mainland France, their origins and introduction vectors from a taxonomic point of view. It focused only on multi-cellular species, meaning that it effectively excluded the protists responsible for parasitic diseases such as *Bonamia ostreae*, which originated in North America and is responsible for the bonamiosis disease that caused the collapse of production of the European flat oyster (*Ostrea edulis*) in the 1980s, and the *Haplosporidium* of the same origin (*Haplosporidium nelsoni* and *H. costale*) that infect Pacific oysters. These parasites have been identified by the World Organisation for Animal Health ([WOAH], formerly the Office International des Epizooties [OIE][5]) as emerging diseases, some of which are notifiable for member countries. Similarly, phytoplanktonic species are not listed in this summary, mainly because of uncertainties over their geographical areas of origin and wide distribution. Some, however, are well known for their exotic nature, such as ostreopsis, which was identified in France a few years ago, particularly along the Mediterranean coast, and which can cause public health

---
5. https://web.oie.int/downld/PROC2020/F_RAPPORT_ANNUEL.pdf

problems due to its ability to synthesise ovatoxins: toxins similar to palytoxin (Lassus *et al.*, 2016).

For all these reasons, the current situation in the French Exclusive Economic Zone (EEZ) is clearly underestimated.

Nevertheless, more than 342 alien species have been identified on the coasts of France that were introduced from the 12th century up to September 2022. Specific inventories have been made since the 19th century, giving a high level of confidence in their description as aliens. The majority (117) come from the temperate North Pacific, and 42 are cryptogenic. Since September 2022, a dozen additional cases of introduction have been reported, either concerning new species or extension to several coastlines. The vast majority (68%) were introduced accidentally via biofouling of ship hulls (37%), ballast tanks (29%), or were associated with animal transport. A third resulted from multiple introductions; for example, up to four routes and vectors of introduction are listed for the filamentous brown alga *Chrysonephos lewisii*. A significant acceleration has been noted since the 1970s, in correlation with the increase in human activity and the process of globalisation. It is also likely that the increase in reports over the last two decades is due to a heightened awareness of the impact of biological invasions, and the resulting paradigm shift, with more stringent regulations and a greater number of dedicated research projects.

In this mainland France inventory, arthropods are the most numerous (69 species), followed by red algae (67), molluscs (44), chordates (43) and annelids or marine worms (38). Most of the diversity of marine phyla is represented, with the exception of echinoderms (e.g. sea urchins, starfish), for which the reason is not apparent. It should be noted that some phyla receive more scientific attention simply because more specialists study them, which can introduce bias into the final assessment. The decline in the community of scientific taxonomists means that some phyla are becoming 'orphans' with little associated scientific expertise, a paradox at a time when the need for knowledge about biodiversity is at the top of the international agenda. This is what the CBD calls the 'taxonomic handicap' (Faugère and Mauz-Arpin, 2013). Another notable feature is that several marine species have been identified as being among the 100 most invasive worldwide (100 worst invasive species[6]), such as common cordgrass (*Sporobolus anglicus*), *Caulerpa taxifolia* and wakame, the Chinese mitten crab (*Eriocheir sinensis*) and American ctenophore (*Mnemiopsis leidyi*), a pelagic species (in overseas waters, the European green shore crab (*Carcinus maenas*) is also on this list [GISD, 2024]).

Some species are particularly invasive, such as *Caulerpa* sp., Japanese wireweed (*Sargassum muticum*), wakame, American blue crab (*Callinectes sapidus*), Asian shore crab (*Hemigrapsus sanguineus*) and brush-clawed shore crab (*H. takanoi*), the urchin crab (*Percnon gibbesi*), common slipper limpet (*Crepidula fornicata*), also from North America, and the Pacific oyster. An exotic seaweed, for example, can cover an area, limiting the penetration of light necessary for the development of native species. Several such species present an invasive risk given the knowledge already acquired about them on an international scale, with invasive processes in progress (e.g. the brown alga *Rugulopteryx okamurae*).

Although the coast of mainland France is divided into distinct ecoregions, almost a third of the inventory is present on all three main coastlines, and 42% are present on two. This potentially represents cases of multiple introductions per species and/or secondary introductions from an initial site. The Mediterranean coast has the highest number of introductions of exotic species (240): the warm waters, ongoing tropicalisation of the Mediterranean Sea and its multiple shipping lanes, due in particular to the Suez Canal, facilitate these introductions. Since 2005, the European DAISIE programme has analysed the situation in Europe and found that an exotic species is introduced into the Mediterranean Sea every six weeks. This was two and a half years for ecosystems in the North Atlantic (Ireland[7]). This information is also available in the European Alien Species Information Network (EASIN)[8] database (JRC, 2024).

---

6. https://www.iucngisd.org/gisd/100_worst.php

7. https://www.gbif.org/fr/dataset/39f36f10-559b-427f-8c86-2d28afff68ca; http://www.europe-aliens.org/

8. https://easin.jrc.ec.europa.eu/easin

HOW ARE EXOTIC MARINE SPECIES INTRODUCED?

A HISTORICAL PERSPECTIVE ON SPECIES INTRODUCTIONS

Throughout history, maritime transport has played a fundamental role in the movement of people and goods. Little is known about the transfer of species associated with the first transoceanic voyages, notably via the biofouling of ship hulls, the woodworm present in their wood and the ballast needed to keep ships stable. Some cases have been identified retrospectively thanks to comparative genetic analyses of populations. The case of the Portuguese oyster (*Magallana angulata*) is emblematic: commercially exploited in European shellfish farming since the 19th century, genetic analyses have made it possible to identify its origin in the waters of the north-west Pacific (Taiwan) and to link its presence in Europe to the great voyages and maritime explorations made by the Portuguese between the beginning of the 15th century and the middle of the 16th century (Huvet *et al.*, 2000). The case of the soft-shell clam is also considered to be an ancient invasion of Europe associated with Viking movements between North America and Europe from the 11th to 14th centuries (Essink and Oost, 2019). Similarly, northern hemisphere mussel species *Mytilus* spp. were introduced to South America as early as the 1500s by the ships of European explorers (Carlton, 1999; Ojaveer *et al.*, 2018).

These transfers and biological invasions have therefore been going on for a very long time, centuries before biologists started making fauna and flora inventories to document species biogeography. It has also been shown that the practice of ballasting ships was already in use in the Bronze Age.

As early as the 1950s, scientists were retrospectively identifying the terrestrial and aquatic fauna exchanged between Europe and North America (Lindroth, 1957). In 1958, Charles Elton published a book entitled *Ecology of Animal and Plant Invasions*. Lindroth (1957) had already identified Newfoundland as the area with the highest incidence of accidental introductions of European species into North America. Historically, the Vikings reached this territory as early as the year 1000, forming permanent settlements for a few years. In 1497, John Cabot rediscovered the island of Newfoundland, which was to become a source of trade for centuries to come, with permanent settlements linked primarily to south-west England.

From the 15th century onwards, intercontinental navigation developed. Ships became larger and stronger, and a round trip from Lorient to India now took just 18 to 24 months. However, the number of potentially invasive species being transferred from one continent to another has accelerated considerably since the end of the 19th century, in line with the development of naval architecture and intercontinental transport.

At the same time, fishing resources were already being heavily exploited in the Middle Ages, particularly by French and Spanish fishing vessels. These fisheries required harbours and access to land via ports that were very busy in summer, but uninhabited in winter. The first permanent settlements of this kind began in the 1600s, and around 1670 for the archipelago of Saint Pierre and Miquelon.

The first regulations on maritime traffic and deballasting can be credited to the founder of the first permanent colony in Newfoundland, John Guy, who imposed eight rules on fishermen, the very first of which was "that no ballast press stones or anything else hurtful to the harbours be thrown out to the prejudice of the said harbours, but that it shall be carried ashore and laid where it may not do any annoyance under the pain of five pounds for every offence".

This desire to preserve port infrastructures was characteristic of the times in Europe. As port maintenance and cleaning work was very costly, it was essential not to make matters worse by unloading ships. In France, the Michau Code (1629) provided a framework for this practice, condemning offenders to the confiscation of their ships, an order that was followed by a succession of regulations over the course of the 17th century. Colbert's 1681 Marine Ordinance provided a technical framework for this practice in France. In the end, it led to the accidental introduction of numerous species of terrestrial fauna and flora (seeds, fruit, insects and species associated with soils) via the deballasting of stones, sand, gravel, etc., which was carried out frequently on both sides of the Atlantic over the course of the 17th and 18th centuries (Llinares *et al.*, 2018).

Although the practices of the period ultimately prioritised the introduction of terrestrial species, marine species were introduced via the biofouling of boat hulls, even though the slowness of these ships' movements limited their survival. Such biofouling nevertheless represented a significant vector of introduction. Researchers took the opportunity of a two-month, 800 km voyage along the west coast of the United States in 1988 on a replica of the *Golden Hinde* to study the survival of communities living on the ship's hull in conditions similar to those in the 16th century (Carlton and Hodder, 1995). The original of this English three-masted ship weighing 100 tonnes and 31 m in length is known to have completed a circumnavigation in 1577–1581, commanded by Sir Francis Drake. Nearly thirty species were displaced during the experimental voyage, including benthic species whose transfer resulted from the boat drying out on the mud at low tide. This experiment is a convincing demonstration of the accidental introduction of species via the biofouling of the hulls of boats from this period. An aggravating factor was that sailing boats required regular maintenance and careening. While it was possible to careen hulls in the open sea (heaving down), it was nonetheless very dangerous and led to the loss of many ships (Goulletquer, 2022). Careening was more commonly practised in shallow waters in ports, in areas where the tides swayed, facilitating the introduction and survival of species detached from ship hulls.

Two major innovations were made in maritime transport in the 19th and 20th centuries: first, the transition from sailing to motorised ships equipped with liquid ballast; and second, the containerisation of the 1960s. The globalisation of trade has since greatly increased the use of maritime transport, which now carries 80–90% of consumer goods and has seen an increase of

460% since 1960. As a result, the carrying capacity of container ships rose from 1,300 to almost 4,000 TEUs between 1992 and 2017,[9] and port infrastructures have had to adapt in parallel, as have roads and shipping lanes. The switch from sail to steam and steel-hulled ships drastically reduced journey times: the duration of an Atlantic crossing was cut from 35 to 15 days, thereby improving the survival of associated exotic species.

The container first appeared in the United States in 1956, revolutionising the transport of goods. By 2021, traffic reached 11 billion tonnes, compared with 2 billion in 1970, with a parallel race towards gigantism. In 2021, the world's largest container ship, the *Jacques Saadé*, carried 21,433 containers (220,000 t) during a port call at Le Havre.

The switch from solid to liquid ballast, pumped in to stabilise the ship and then deballasted in the port of arrival, resulted in a drastic change in the nature of the species introduced, initially mainly of terrestrial origin, but now marine. In the 2000s, the International Maritime Organisation (IMO), an agency of the United Nations specialising in issues relating to maritime transport, estimated that 10 billion m^3/year was deballasted in ports, which represents the introduction of more than 1,000 marine species per day. Until 2017, one estimate put this untreated deballasting in French ports at more than 20 million m^3/year.

Maritime transport is considered to be the main vector for the introduction of marine species for three reasons: the presence of liquid ballast, the presence of biofouling on the hulls of ships and the creation of new maritime routes (Asia–Europe, development of maritime hubs) and sea lanes (Suez Canal, Panama Canal); these canals are responsible for breaking down natural biogeographical barriers.

Nevertheless, scientific observations directly linked to introductions via ballast began to be documented at the end of the 18th century, in line with the associated technological developments: the introduction of the rough periwinkle (*Littorina saxatilis*) from western Europe to the Adriatic Sea (1792) and the common

---
9. TEU = size of a 20-foot equivalent unit container.

periwinkle (*L. littorea*) to the North American coast (1840s) are two examples resulting from the use of solid ballast. Another example is the detection of smooth cordgrass, *Sporobolus alterniflorus* (Loisel.), a halophytic plant that arrived in France from North America at the beginning of the 19th century (1803) and still thrives today on the salt marshes of Brittany. The European green shore crab, which was introduced to the west and east coasts of North America in 1817 via solid ballast and marine biofouling, remains one of the major predators on the American and Canadian coasts today (Edgell and Hollander, 2011). This species has also recently arrived on the coasts of Saint Pierre and Miquelon, where it is proliferating with significant impacts on marine habitats (seagrass beds and lagoons) (Sellier *et al.*, 2016).

CURRENT INTRODUCTION VECTORS

Present vectors of introduction correspond to the physical means and/or mode by which the species are introduced. Examples include ballast water from merchant ships, the hulls of boats on which marine biofouling can be found, and international trade in seafood products.

A distinction can therefore be made between voluntary introductions, such as the introduction of a species for aquaculture purposes, and accidental introductions, such as during the deballasting of commercial vessels.

The number and diversity of introduction vectors have evolved and increased over the centuries and are continuing to diversify today. In 1800, only two mechanisms were identified: biofouling of the wooden hulls of sailing boats and the solid ballast used to stabilise them. At the end of the 19th century, three additional vectors were added: ballast water from steel-hulled ships, international trade, and imports of species for aquaculture purposes. By 2000, several dozen more vectors of varying importance linked to human activity could also have been contributing to these introductions. These ranged from several cases of the introduction of marine species, anecdotally associated with the use of seaplanes, to major vectors such as maritime transport. The proliferation of maritime infrastructures (ports, jetties, offshore wind farms) facilitates the establishment of exotic species through the increased availability of artificial habitats, while modifying the connectivity of different areas. The increase in intensity specific to each vector must be taken into account (e.g. increase in maritime transport), as must the specific nature of vector-species pairs, which may favour certain categories or groups of species. For example, until recently, the 'ballast water' vector facilitated the introduction of pelagic species and those whose reproduction is characterised by a pelagic larval phase.

Nowadays, the introduction of exotic species is increasingly considered to be multimodal, i.e. the result of multiple episodes and routes of introduction. The fact that a species can be introduced by a number of different means makes management, particularly in terms of prevention, all the more complex.

Given the possible number of vectors and routes of intentional or accidental introduction, it is necessary to categorise them. The following are the main categories.

Boats, mobile platforms and other means of navigation

Maritime transport has long been involved in the introduction of a number of exotic species into port infrastructures and main shipping lanes. These species may be pelagic and/or sessile, with fixed or associated organisms (epibionts) as well as potential parasites and pathogens. Microorganisms are particularly concerned and have probably been the source of unexplained emerging diseases on a global scale. This is currently the case with the disease affecting sea urchin populations in several marine ecoregions, first detected in early 2022. The disease caused a significant increase in mortality among long-spined sea urchins (*Diadema antillarum*), first in the Caribbean, then in a few months arrived in the Mediterranean, Red Sea and Indian Ocean (e.g. Réunion Island), where it continued to decimate sea urchins. The agent responsible is *Philaster apodigitiformis*, a single-celled ciliate with vibrating surface cilia, already recognised as a threat to sharks, other fish, and crustaceans. Its spread, facilitated by maritime transport, caused this mass mortality. The rapid arrival of this ciliate in the Indian Ocean poses a threat to

Australia, particularly the Great Barrier Reef, which has already been weakened by bleaching episodes caused by marine heat waves. The health of coral reefs is directly linked to that of sea urchins, as they control populations of algae that can suffocate corals through their proliferation.

Opportunities for the introduction of non-native species through maritime transport have increased in recent decades, with the concomitant growth in port infrastructures and their multiple uses. Paradoxically, the improved quality of port waters has made it easier for these introductions to survive.

Ballast water and sediment

The use of liquid ballast began in the 1870s. Depending on the type of ship, water, generally taken from ports, is pumped into individual tanks and distributed along its length. The tanks are managed individually or collectively, depending on the cargo level, to maintain the ship's stability and can hold up to 30% of the cargo load capacity. These volumes of water and sediment in suspension can have heterogeneous characteristics within the same vessel. One cubic metre can contain up to 50,000 individual zooplankton and 110 million phytoplankton cells, enough to form the basis for subsequent populations. For these reasons, ballast water is considered to be the primary vector for the introduction of alien marine species, accounting for 60% of documented cases. Deballasting on the high seas to prevent the introduction of species into ports has long been a recommendation of the International Maritime Organisation (IMO), with a requirement for three ballast renewals per voyage (purging around 95%). However, such movements of water masses at sea can induce structural stresses and affect ship hulls. In addition, heavy sea conditions can make these operations dangerous for the stability of the vessel. Although necessary, these recommendations make the operation technically uncertain and risky. More recently, the implementation of the IMO Ballast Water Management (BWM) Convention, finalised in 2017 and fully operational in September 2024,[10] completes

---

10. https://www.imo.org/en/MediaCentre/HotTopics/Pages/Implementing-the-BWM-Convention.aspx

the system by requiring systematic ballast water treatment. This is one of the most significant advances in preserving the marine environment made in recent years.

Biofouling

When immersed for a long time, a boat hull quickly becomes an 'artificial reef' sheltering a number of fixed and mobile species, sometimes in several successive layers. Wooden-hulled boats can also be affected by the colonisation of wood-eating species (xylophagous). As they move, boats carry such biofouling, which are still a major vector for the introduction of species. Antifouling paints are generally applied at regular intervals to limit the attachment of species and slow the oxidation of metal hulls. For a merchant ship sailing from one continent to another, biofouling can add up to 10% to the weight of the ship and hinder navigation, encouraging the use of antifouling/biocide paints (also sources of pollutants) that alter bacterial films and limit the organism attachment. However, imperfections and wear marks on hulls or cavities (water intakes, housings) facilitate the development or survival of species. For instance, gametophytes of wakame, one of the most invasive marine species in the world, only need microcavities to survive and can withstand long dry spells, making it easier for them to disperse (Epstein and Smale, 2017).

There is today a general framework established by the UN Convention on the Law of the Sea, which calls on States to prevent, reduce and control pollution caused by man in the marine environment, in particular the deliberate or accidental introduction of foreign species into a particular ecosystem. However, measures to limit accidental introductions via biofouling of ship hulls are not based on regulations, but on (non-binding) recommendations such as those issued by the IMO. Only a few countries are stricter in this respect, such as New Zealand, Australia, Brazil and Norway, and can turn away vessels that do not have an up to date 'cleanliness certificate'. However, the regulatory framework is gradually evolving for this vector of introduction, which currently represents the priority to be addressed: in 2023, the IMO adopted a resolution establishing guidelines for managing biofouling and minimising the risk of transferring invasive species. The practical arrangements for management have not yet been established, despite the proposal by organisations such as the ship owning association Bimco and the International Chamber of Shipping (ICS) to the IMO to apply new operational procedures: criteria for choosing the cleaning operator (their certification), crew training, ship preparation, environmental and safety requirements, and post-cleaning inspection.

International trade (import–export)

International trade in seafood products, which in many cases require re-immersion in an open or semi-open environment before sale, has been growing since the 19th century. Originally from the east coast of the United States, the Eastern oyster (*Crassostrea virginica*) was imported into Europe in the 19th century and was the initial vector for the introduction of the common slipper limpet, an invasive gastropod mollusc. This shellfish, which has greatly modified the marine environment, is now in trophic competition with farmed shellfish, while remaining invasive on the French coast a century later. Similarly, American lobsters have been found on European coasts following escapes from storage tanks, and the veined rapa whelk (*Rapana venosa*), an Asian gastropod that predates farmed shellfish, was found on the Atlantic coast following commercial flows of shellfish from Italy (ICES, 2004).

Leisure activities

Many leisure activities can potentially be vectors of primary and/or secondary introductions of exotic species. In the latter case, unmaintained fishing gear or pleasure craft may have facilitated their dispersal. Recreational boating is considered to be a major vector of both primary introduction and secondary spread due to the number, distribution and connectivity induced by the concentrations of these boats in ports. The number of marinas has risen sharply since the 1960s. By way of example, around 1.5 million pleasure boats were counted in the Mediterranean Sea by satellite imagery in 2007.

Several primary introductions are associated with international trade for aquarium and leisure activities, such as recreational fishing, which has become very popular. Available from pet shops or directly from the internet, non-native species include fish, algae and invertebrates. This international trade can be a source of accidental introductions of potentially invasive non-native species. For example, the live bait sold for recreational fishing mainly consists of marine worms originating from North America and Asia (Sa *et al.*, 2017; Font *et al.*, 2018). With recreational fishing booming, the collection of marine worms from the local natural environment is no longer sufficient to support this live bait market. The import market for polychaete and sipunculid marine worms has therefore developed to meet this demand, as well as aquaculture to produce them (Pombo *et al.*, 2020). It is now very easy to obtain these bait species online and have them delivered live in less than 24 hours! Moreover, they are packaged using seaweed, sand and other materials likely to harbour additional exotic species. If recreational fishers are not informed of the environmental risks of using this bait, the likelihood of accidental or deliberate introductions into the marine environment increases, potentially undermining biodiversity conservation efforts. For example, Arias *et al.* (2013) highlight the presence of perennial reproductive populations of the Korean polychaete *Perinereis vancaurica*, introduced via this bait trade into the Mar Menor lagoon in Spain, and warn of the associated ecological consequences due to its invasive nature.

The deliberate introduction of marine species for ornamental purposes goes back a long way. As early as 1866, the American horseshoe crab (*Limulus polyphemus*), a kind of 'living fossil', was imported from the United States to Europe, without any permanent populations developing. Nowadays, large public aquariums have infrastructures that limit the risks of escape by treating effluent. However, there are two emblematic examples linked to this activity. Firstly, the accidental escape of *Caulerpa taxifolia* into the Mediterranean Sea in 1984, followed by a massive biological invasion until 2010 and affecting the Mediterranean ecosystem of several countries during this period. In particular, this species has come into competition with Neptune grass (*Posidonia oceanica*), a plant endemic to the Mediterranean that occupies between 20% and 50% of the coastal seabed at depths of between 0 and 50 metres. Neptune grass meadows host more than 20% of Mediterranean biodiversity, making them a key priority for the protection and management of the region's marine environment (Boudouresque *et al.*, 2012; Pastor *et al.*, 2023).

The second case study concerns the first reports of red or common lionfish, in Florida in 1985, as a result of repeated escapes from aquariums. A formidable predator that attacks coral reefs, *P. volitans* was considered established in the 2000s. This fish is recognised neither as prey nor as a predator in its new ecosystem, hence its invasive success! It colonised the Greater Caribbean and Gulf of Mexico in less than a decade. It has currently spread as far south as the Brazilian coast and as far north as Cape Hatteras in North Carolina, where the cold waters (below 16°C) seem to be limiting its expansion.

Other vectors

More recently, floating waste has been identified as a significant vector of species introductions. According to UNEP (2021), plastics account for over 85% of waste at sea. As well as causing direct pollution, this plastic provides a favourable habitat for many species and is transported by marine currents. A review of the scientific literature indicated 387 taxa associated with floating debris (Kiessling *et al.*, 2015). With their persistence in the environment and good buoyancy, these plastics allow species to be dispersed much more widely than by natural processes and modify the connectivity between different ecosystems (Maes, 2022; Mghili *et al.*, 2023). A reassessment of the respective importance of each vector of introduction has highlighted not only their significant contribution, of the order of 5% of waste at sea (e.g. macro and microplastics), to this process (García-Gómez *et al.*, 2021b), but also the fact that this waste is both a primary and secondary vector of introduction, facilitating the dispersal of IAS (Mghili *et al.*, 2023).

It should also be noted that catastrophic events can contribute to the process of introducing exotic species. The Tōhoku earthquake in March 2011, which triggered a deadly tsunami and the Fukushima industrial accident, was responsible for 5 million tonnes of waste at sea, 47% of which was plastic. This release of waste resulted in massive arrivals of waste and invasive species on the west coast of North America as early as 2012, after a journey of 4,000 nautical miles (Therriault *et al.*, 2018). These spring arrivals have been documented for more than six years, with the prospect of more new introductions for more than a decade. Finally, an entire ecosystem landed on the American coast. A floating pontoon alone was home to more than 100 species of Asian origin, many of which were known to be invasive, such as the pancake batter tunicate (*Didemnum vexillum*), an ascidian, or the blood crab, but also species already known as 'alien' in Asia, such as the Mediterranean mussel and the common starfish (*Asterias rubens*). More than 280 species were recorded on 600 different types of waste (including plastics), mainly invertebrates but also two fish (Craig *et al.*, 2018; Tan *et al.*, 2018).

CURRENT ROUTES OF INTRODUCTION

Introduction pathways correspond both to the reasons for an introduction, for example the voluntary introduction of a species for aquaculture purposes, and to the practical and geographical way in which this introduction takes place (e.g. passage through the Suez Canal).

Development of aquaculture

The significant development of aquaculture since the 1950s has greatly increased the demand for seafood products and plays an important role in addressing the global food challenge (FAO, 2024). The various forecasting exercises in this area point to an increase in aquaculture production over the coming years in the face of stable fisheries tonnages since the mid-1990s, a decline that is foreseen to continue into the future. Aquaculture accounted for 46% of global production of aquatic products in 2018 (Goulletquer and Lacroix, 2022). In 2022, aquaculture production overtook fisheries production for the first time, reaching 51%, and has been growing at an annual rate of 6.6% since 2020, reflecting the sector's dynamic growth. An analysis of aquaculture statistics highlights the importance of exotic species in the economic performance of these production sectors (FAO, 2024).

A large number of species have been introduced worldwide so that aquaculture can benefit from their superior performance compared with native species, their tolerance of various temperatures and salinities, the absence of competitors, predators and pathogens, their ease of rearing at high densities, reproduction methods and commercial interest. White-leg shrimp (*Penaeus vannamei*), originating from the Pacific coast of Latin America, were imported en masse into various countries in the 1970s and now account for 76% of the world's penaeid production. Some species, such as the Pacific oyster and the Manila clam, are now farmed on several continents and are part of the globalisation process. The Pacific oyster, for example, has been introduced to more than 64 countries and 10 overseas territories. It has subsequently developed naturally in 32 countries and is the subject of aquaculture production in 36 (Martínez-García *et al.*, 2021). Many invertebrates are reared in aquaculture worldwide, but there are also emblematic fish species (e.g. Atlantic salmon) and cultivated algae (e.g. wakame) (Goulletquer, 2016; Epstein and Smale, 2017). For several of these species, these introductions have led to massive dispersals and the development of feral populations, posing difficulties for the receiving ecosystems, but also sometimes providing a source of new fisheries. French production is based on several exotic species, such as the Pacific oyster and the Manila clam, hard clam, kuruma shrimp or blue shrimp (*Litopenaeus stylirostris*) in New Caledonia (originally from Central America) and the red drum (*Sciaenops ocellatus*) in several overseas territories. For the oldest introductions, only human and animal health conditions were considered in the practical arrangements, without any real consideration of the environmental impact of their potential invasive nature. The problem of biological invasions had not yet been assessed in its entirety, leading to relatively limited precautionary measures and the accidental introduction of many associated species, including pathogens (Grizel and Héral, 1991; Gruet *et al.*, 1976). The case study of French oyster farming will be covered in more detail later.

Approaches to populating geographical areas in order to develop new fisheries have driven the dispersal of species, sometimes to the point of invasion. The Kamchatka crab (*Paralithodes camtschaticus*), which comes from eastern Siberia, is highly prized worldwide and can weigh over 12 kg. It was deliberately introduced by the Russians in the 1960s into the Bay of Kola in order to develop a new fishery. With no predators, the species quickly spread to the Barents Sea where it affected ecosystem function. Despite management measures, the species continues to spread southwards. Similarly, the pink salmon (*Oncorhynchus gorbuscha*) is native to the Pacific Ocean. It was also deliberately introduced in the 1950s into the Kola Peninsula in northern Europe, and accidentally into the Arctic and North America as far as the American Great Lakes. An alert has now been issued by the North Atlantic Salmon Conservation Organisation (NASCO) due to its rapid expansion and impact on wild populations of native species, particularly Atlantic salmon, classified as near-threatened since 2023 on a global scale according to IUCN criteria. The species has now been reported as far as the rias of Northern Brittany (NASCO, 2022a; 2022b; 2023).

Waterways and maritime infrastructures

The globalisation of trade in goods is directly correlated with maritime routes and corridors, which will be used increasingly in the future because maritime trade is growing by 3.5% annually. At present, more than 90% of trade in goods passes by these routes.

While new sea routes, such as the Northern Sea Route, or Northwest Passage, with potential new species introductions, are opening up as a result of the effects of climate change, canals remain a major route of introduction on a global scale. International conventions such as the CBD and United Nations Convention on the Law of the Sea (UNCLOS) recommend that canals be dealt with in a specific way, given their contribution to biological invasions and biodiversity loss. Canals have transformed maritime traffic since at least the 6th century BC, when the first navigable canal linked the Mediterranean Sea to the Red Sea via the Nile.

Natural geographical barriers used to structure the distribution of species, limiting their dispersal. The deepening of these inter-oceanic canals represents a break in these barriers, providing new opportunities for marine flora and fauna to disperse, both passively and via maritime transport. New connections have emerged where ecosystems had been separated for millions of years. Today, the Suez and Panama canals are considered to be major routes for the introduction of exotic species, some of which are invasive (Galil *et al.*, 2018; Castellanos-Galindo *et al.*, 2020).

The Suez Canal

Even before it was dug, scientists saw the Suez Canal as an opportunity, rather than a risk, to study the dispersal of species and the mixtures that would result. The Suez Canal has been operational since 1869. In less than a decade, two species of bivalves originating from the Red Sea had already been documented in the Mediterranean Sea: the Atlantic pearl oyster (*Pinctada imbricata radiata*) and Red Sea mussel (*Brachidontes pharaonis*). By 2015, 443 species of macroalgae, invertebrates and fish had been introduced via the Suez Canal, 89 of them to more than five Mediterranean countries (Galil *et al.*, 2018). In 2022, 90% of the 464 exotic species identified on the Israeli coasts of the Levant were there because of the Suez Canal. Only a few species have made the opposite journey to the Red Sea, these include the starfish *Sphaerodiscus placenta*, the white-speckled headshield slug (*Biuve fulvipunctata*), (Vitale, 2017), the peacock blenny (*Salaria pavo*) and the Egyptian sole (*Solea aegyptiaca*) (Chanet *et al.*, 2012).

The Suez Canal is the shortest route between Europe and Asia, offering an alternative to the 9,000-km sea route around the African continent. This immense work, inaugurated in 1869 and subsequently enlarged and modernised several times, is an indispensable source of revenue for the Egyptian state budget. The Suez Canal Authority (SCA) announced that it would reach its highest level of turnover in 2022 with $8 billion, an increase of 25% compared with 2021. Over the period 2016–2022, canal revenues totalled $41.7 billion, compared with $35.4 billion over the period 2008–2014 (an increase of 18%), despite the internationally troubled period caused by the Covid-19 pandemic and Russian-Ukrainian war. Passage fees have risen by between 10% and 15%, depending on the vessel. In 2022, this canal linking the Mediterranean to the Red Sea saw 24,000 ships pass through, carrying 1.4 billion tonnes of goods: an average of 68 ships a day (56 in 2021, 47 in 2014). Between 10% and 12% of world maritime trade passes through the Suez Canal. The scenario of a halt to traffic on the canal is feared by world markets. The stranding of the giant container ship *Ever Given* for several days in 2021 due to strong winds blocked the canal for six days while the ship was being rescued. It resulted in the death of an SCA agent and an estimated financial loss of between $12 million and $15 million per day of closure. The ship, which was travelling in excess of the 12-knot speed limit, lost control and ran aground. Consequently, 400 ships were delayed and insurers estimated the daily loss to world shipping at several billion. This incident served as a reminder of the profound dependence of the global economy on maritime transport.

At the same time, the construction of 'maxi' container ships (>12,000 TEU and, with the largest being >19,000 TEU) requires the expansion of port infrastructures and shipping lanes. On 30 June 2022, Evergreen's *Ever Art* (23,992 TEU), the world's largest container ship, entered the canal for the first time.

When the Suez Canal opened in 1869, it was 8 m deep. Gradually, it was widened, then dug deeper, and its five sections were doubled. The initial cross-section had an area of 304 m^2, which was then increased to 1,200 m^2 in 1956 (14 m), 1,800 m^2 in 1962 (15.5 m), 3,600 m^2 (19.5 m) in 1980, and 5,200 m^2 in 2010 (24 m). In 2015, scientists alerted the public to an extension project carried out without an environmental impact assessment, doubling the canal by adding a 72.4 km corridor parallel to the existing one and significantly increasing the potential for introducing exotic species to the Mediterranean (Galil *et al.*, 2015). Such developments require impact studies based on the latest scientific knowledge. Signatories of the CBD have an explicit obligation in this respect, also considering transboundary impacts on biodiversity and IAS. Various international conventions require states to ensure that their activities do not harm the environment of other countries. An alert was issued to international bodies such as the CBD and the UNEP Barcelona Convention on the impacts of the project, drawing attention to established scientific knowledge on the effects of IAS, while calling for a regional approach to this development and a scientifically based risk assessment (Environmental Impact Assessment: EIA).

EIAs enable impact mitigation measures to be drawn up sequentially (ARC sequence: avoid-reduce-compensate).

Ultimately, the digging of the new section of the Suez Canal in 2015 has shortened crossing times and made it easier for ships to pass each other. This increase in both width and depth has led to the introduction of a cohort of new potentially invasive exotic species into the Mediterranean Sea. Since 2015, eight new Lessepsian fish species have become established in the Mediterranean, representing an 8% increase in species and a doubling of the annual detection rate (compared with 1869-2015). In three years (2019-2021), 72 new exotic species were detected in the Mediterranean Sea.

From an environmental point of view, there are significant differences between the Red Sea and the Mediterranean (e.g. salinity of 39 g/l vs. 30 g/l). The ongoing 'tropicalisation' of the Mediterranean tends to reduce these differences, particularly in temperature, and facilitates the expansion of species' ranges towards the west and north-west of the Mediterranean. For a long time, it was thought that only (Lessepsian) species adapted to and confined to shallow depths could be introduced in this way because of the shallowness of the canal. However, the recent enlargements have increased the volume of seawater and current speeds, making it easier for propagules of deeper species to spread. Moreover, the increase in temperature in the Mediterranean Sea is improving their survival rate. For example, crabs and fish from the Red Sea have been found at depths of 250 m off the Israeli coast. Red Sea species are also highly adaptable due to their original environmental conditions, making them formidable competitors with Mediterranean species. Some of these species, such as the Red Sea mussel (*Brachidontes pharaonis*), rivulated rabbitfish (*Siganus rivulatus*) and square-tailed rabbitfish (*Siganus luridus*), bluespotted cornetfish (*Fistularia commersonii*) and silver-cheeked toadfish (*Lagocephalus sceleratus*) have already reached the French coast, while others are expected to follow the expansion of their geographical distribution from east to west, stimulated by the current warming of Mediterranean waters. These observations have shown that the expansion and invasiveness of certain species have been underestimated.

The Egyptian authorities are looking for ways to limit the invasion of marine life from the Suez Canal. The Bitter Lakes are the most significant bodies of water in the canal, accounting for 85% of the volume of water in almost 24% of the canal's length. Oligotrophic in nature, significant changes in their thermal and hydrological regimes have been noted in recent decades (El-Serehy *et al.*, 2018). Historically, the high salinity of the Bitter Lakes, located in the middle of the canal, represented a hypersaline barrier for many species, restricting their spread towards the Mediterranean. However, this barrier has been disappearing in recent years: the 276 M m^3 of agricultural wastewater discharged into the Bitter Lakes every year has considerably diluted their salinity levels and rendered them ineffective as a natural barrier.

Preventive methods have been considered by the Egyptian government. In an environmental impact study that has not been made public, the authorities studied the possibility of establishing a 'bubble curtain', in which air is injected into underwater pipes pierced with tiny holes to create turbulence in order to deter fish. Another approach involving broadcasting sound aimed at fish species has also been considered. An eco-engineering approach using brine from desalination plants could increase the salinity of the Bitter Lakes. Restoring a saline barrier by redirecting wastewater into other channels remains a priority option. The salinity shock thus created is considered to be the best management option.

The Panama Canal

The Panama Canal, operational since 1914, links the Pacific and Atlantic oceans via the Isthmus of Panama (80 km) in Central America. Before it was built, and became a vital link for maritime trade, ships had to pass around Cape Horn and through the Drake Passage (at the tip of South America) to reach either of the oceans now linked by the canal. To reach San Francisco from New York, a ship had to travel more than 22,000 km via Cape Horn. It now needs only to travel 9,500 km via the canal, a considerable time saving. The Panama Canal is a strategic crossing point for shipping, with an annual average of 203 million tonnes of cargo passing through it in the 2000s. From its construction until 2002, more than 800,000 ships passed through the canal.

In fact, 40% of goods transiting between north-east Asia and the east coast of the United States use it.

Unlike the Suez, the Panama Canal contains freshwater. It lies about 29 m above sea level and has locks at both ends. These locks are gravity-fed, with associated basins to optimise water management and reduce water losses by half. To pass through a lock, around 200 million litres of fresh water need to be discharged, which is obtained from the hydrographic basin combining the Gatún and Alajuela lakes. Hydraulic management is precise but depends directly on rainfall levels in the area to supply Lake Gatún in particular. The canal's catchment area is also a source of drinking water for Panama City and Colón, and for many villages between these two ends. Half of the country's 4.2 million inhabitants obtain their drinking water from this basin, which is currently suffering from a cruel lack of rainfall, leading to conflicts over the use of freshwater resources.

Gatún, a large artificial freshwater lake located between the lock systems, formed a 'natural' barrier limiting the passage of marine species from the Caribbean to the eastern Pacific. However, since 2007, recent extensions to the canal have reduced this barrier capacity. The widening work opened the way to ships capable of carrying more than double (12,000 containers) the load previously authorised to pass through the canal. The third lock level was built specifically for the larger 'Post-Panamax' ships. The first of these ships, *Neo-Panamax*, passed through the new locks on 26 June 2016. By 2022, 14,000 ships carrying 518 million tonnes of cargo had passed through the canal. However, this design of the third lock system appears to be more conducive to fish passage, operating in a similar way to a fish pass.

Recurrent droughts, due to the lack of rainfall, affect the entire catchment area, causing lake levels to fall. This causes severe disruption to shipping traffic for periods of several weeks. The two artificial lakes of Alajuela and Gatún may soon run dry by April 2023, forcing the authorities to restrict access to this route linking the Atlantic and Pacific Oceans, through which 6% of the world's maritime traffic passes. The repercussions of climate change are a direct threat to international economic interests.

Already in 2019, the canal would only hold 3 billion m³ of fresh water, whereas 5.2 billion m³ are required for normal operation. Although specially designed to maximise the volume of cargo passing through the canal, the largest container ships ('Post-Panamax' class) could no longer pass through.

With successive droughts, the water level of Lake Gatún has fallen sharply, as has that of the Panama Canal, whose salinity is increasing, reducing the gap with strictly marine waters. This trend is a major factor in the restructuring of the lake's pelagic communities. The abundance of species has increased to such an extent that some freshwater fish species have been virtually replaced. Several hundred species of fish can develop in brackish water and potentially migrate beyond the canal. In the century prior to the 2016 development, 18 species of marine fish were sampled in Lake Gatún. The oldest and best-documented species is the Atlantic tarpon (*Megalops atlanticus*), reported as early as 1935 in this lake, and again in 2011 in a lagoon on the Pacific coast of Costa Rica. Between 2019 and 2020, 11 new species were recorded. The risk of red lionfish migrating to the eastern Pacific should not be overlooked. Known to be tolerant of low salinity levels, the red lionfish has already had a major impact in the Caribbean as a highly active predator, with a notable invasive character.

In terms of management responses, in addition to the toll for access to the canal, the authorities had to (re)take drastic measures between July and August 2023, as they had already done during the droughts of 2019 and 2020. Vessels with excessively deep draughts (the height of the immersed part of the boat) (threshold at 44 feet or 13.4 m) were refused entry because of the risk of them running aground. An additional tax on the freshwater used to operate the locks was also introduced. Daily passage was limited to 32 ships instead of 40. The largest container ships had to partially unload their cargo for transport overland between the port of Panama and the port of Colón in the north, where they were reloaded. For example, the maxi-ship *Ever Max* (Evergreen), weighing over 165,000 tonnes, had to unload 700 of the 7,400 containers on board for a rail crossing. Otherwise, it could have set a record for the heaviest load ever to travel through the canal! More than 200 ships found themselves waiting, some for 20 days, to go through the 80 km of canal.

These restrictions led to a 36% increase in shipping prices. The economic losses for 2023 have been estimated at over $200 million in tolls for the authorities. These measures were extended for a year due to the rainfall deficit, in order to facilitate transport planning for shipowners and avoid bottlenecks in shipping traffic. The Panama Canal Authority estimated the tonnage of goods transiting the isthmus at 500 million tonnes, down from 518 million tonnes in 2022. This represents a significant drop in revenue, compared with peak toll revenues of over $3 billion in the past.

An alternative solution to meet growing demand is now being studied: a land link (motorway and rail) linking the ports of Salina Cruz (Pacific) to the Atlantic coast (Coatzacoalcos) via the Isthmus of Tehuantepec.

The Panama Canal Authority is therefore conducting new studies to identify potential water resources. It should be noted that deforestation in the catchment area is an aggravating factor for water management. This is further hampered by the presence of another invasive plant, wild sugar cane (*Saccharum spontaneum*). Taking climate change into account requires new management and adaptation methods to meet the various uses and services provided by freshwater. These management methods can only be applied through integrated approaches in a context of partnership, with the participation of all stakeholders according to the three pillars of sustainable development (social, economic, environmental), while taking into account the different pressures on this environment.

From an environmental and operational point of view, the experiments carried out and methods used to manage IAS in the American Great Lakes — tracking, environmental DNA for monitoring, sonar, electrical management and fish repellent tools (sound, bubble curtain, etc.) — constitute management approaches of interest for preserving Lake Gatún's role as a freshwater barrier to marine species, and for reducing cases of involuntary introductions via this waterway.

WHAT ARE THE IMPACTS OF BIOLOGICAL INVASIONS?

Similarly to the conclusions of the CBD, the global assessment of biological invasions carried out by the IPBES in 2023 confirms that this process is one of the five factors directly responsible for biodiversity loss. In addition to the loss of marine biodiversity, the services provided by nature, human fishing and aquaculture activities, human health and the development of industrial infrastructures are also directly affected. Taking all ecosystems together, biological invasions have contributed to 60% of recent population extinctions, and represent the sole extinction factor in 16% of cases. Island systems are particularly hard hit. It should be noted that, unlike on land, no case of total extinction of a species has been documented in the marine environment as a whole; only populations in specific geographical sectors have suffered extinctions. This resilience at the species level is due as much to the physico-chemical characteristics of the marine environment as to the methods of reproduction and spatial distribution of marine species. However, these processes profoundly modify the functioning of ecosystems and their resilience, as well as the services provided by nature. More than 85% of the impacts are considered to be negative, affecting nature in more than 70% of cases. Back in 2006, the European DAISIE project identified economic losses associated with these introductions at €11.4 billion per year for Europe, broken down into control costs (1.8) and damage costs (9.6). The loss of income for aquaculture and fishing was of the order of €150 million per year. Today, the economic costs incurred are multiplied by four every decade, estimated at a minimum of $423 billion in 2019 on a global scale, the same order of magnitude as the costs resulting from natural disasters (IPBES, 2023; Turbelin *et al.*, 2023). Although few marine case studies have been documented to date, the regularly updated international InvaCost database is useful for assessing the scale of these induced economic costs (Leroy *et al.*, 2022). These figures alone justify management and regulatory approaches aimed at preventing and limiting biological introductions and invasions.

CHARACTERISTICS AND METHODS OF IMPACT ASSESSMENT

Proper assessment of biological invasions requires defining an impact as a measurable change, either in ecological conditions, in the services provided by nature, or in quality of life. The cumulative effect of these impacts, which can potentially act in synergy, must be taken into account, as must their temporal evolution. A distinction can be made between quantifiable changes, for example in the physico-chemical parameters of the environment, and changes in the uses and services produced by the environment, which are intrinsically more subjective in nature and can be considered 'positive' or 'negative' depending on individual perceptions and the spatial and/or temporal scale at which they are viewed. Supply through fishing and aquaculture, for example, can be considered 'positive'. In France, many of the aquaculture species mentioned above have been introduced for production purposes: wakame, Manila clam, Pacific oyster, hard clam, kuruma shrimp, etc. The same applies to the French overseas territories, with blue shrimp in New Caledonia and red drum in the West Indies, Mayotte and Réunion.

Biological invasions can alter environmental characteristics and biological interactions, thereby affecting ecosystem functioning to varying degrees. The term 'ecological impact' (meaning negative and harmful impact) is used if this functioning alters the environment and/or communities through a reduction in the performance of species or a reduction in the populations of native species. Alteration of the environment can result in a loss of ecosystem resilience in the face of other pressures. Conversely, invasion by a species might lead, for example, to opportunities for recreational fishing, which is a source of food and well-being and is perceived as a positive change. An integrated vision of these invasion processes seems necessary in order to assess both the positive and negative effects, in terms of both the environment and the beneficial or harmful effects for humans. For the latter, most of the negative and/or positive impacts have been documented in the marine environment through the prism of supply services (fisheries, aquaculture) (IPBES, 2023; Katsanevakis *et al.*, 2014; Tsirintanis *et al.*, 2022, 2023). It should be noted that the components of quality of life are both tangible and intangible (e.g. cultural services, see **Figure 3**).

Different approaches have been developed to assess the magnitude of such changes: the IUCN (2020) devised the EICAT classification (Environmental Impact Classification of Alien Taxa) to assess the impact on individual performance, populations and local or global extinctions. Five categories have been defined, ranging from minor to massive impacts, to facilitate prioritisation and action. According to the SEICAT (Socio-Economic Impact Classification of Alien Taxa) approach, impacts on quality of life are classified by assessing the level of disruption to human activities resulting from biological invasions, some of which are slowed or even halted (e.g. fisheries) (Bacher *et al.*, 2018). In addition, procedures for assessing the risks associated with the introduction of exotic species are being developed using a global, semi-quantitative approach, enabling current and future impacts to be taken into account in response to climate change (e.g. the Aquatic Species Invasiveness Screening Kit: AS-ISK multilingual application; Vilizzi *et al.*, 2021).

The mechanisms involved are of different kinds. Overall, the effects of a biological invasion will increase in response to the increase in the density of introduced invasive organisms (Shea and Chesson, 2002), as interspecific competition for space and food can lead to the local rarefaction or extinction of populations of native species. Predation can lead to similar results, as can genetic hybridisation between native and exotic species. Several cases of the introduction of parasitic and/or pathogenic taxa have also led to local extinctions of native species. The nematode worm *Anguillicola crassus*, native to Asia, parasitises the European eel (*Anguilla anguilla*), reducing its fertility and survival. This is one of the factors behind the steep decline of the species, which is now considered threatened with extinction. In addition, introduced taxa can be toxic when consumed, with harmful effects. The spatial extension of an exotic species can also have significant physico-chemical consequences, such as a blanket of algae capturing all the light to the detriment of local species.

The local characteristics of an environment can be modified by the development of the alien species (e.g. sedimentation level, granulometry of the seabed), or even by the creation of a new habitat/ecosystem in the case of so-called 'engineer' species, for example the creation of three-dimensional oyster reefs where previously only a featureless mudflat existed. These new reefs can be welcome though as a way of holding back coastal erosion in the face of the effects of climate change (Shakspeare *et al.*, 2024). Such modifications can be seen as a nature-based solution to an environmental problem.[11] Interactions between species, however, whether native or alien, can result in new negative functionalities for the ecosystem, facilitating, for example, the establishment of further alien species; these are known as synergistic effects and 'novel ecosystems' (Simberloff and Von Holle, 1999). In this context, it is necessary to consider the cumulative effect resulting from the impact of each species, all of which may act in synergy (Tsirintanis *et al.*, 2023). These changes may limit the effectiveness of marine protected areas, whose initial purpose is to develop management methods to preserve marine biodiversity.

CASE STUDIES

The following case studies illustrate the nature of these different impacts.

Wakame

Known as 'wakame', the brown seaweed *Undaria pinnatifida* makes an interesting case study. In its native range in Asia, this species is cultivated for medicinal and food purposes (trace elements, high in protein, low in fat), with more than 2 million tonnes produced annually. Native to the north-west Pacific, this species has been the subject of deliberate and accidental

---

11. https://uicn.fr/solutions-fondees-sur-la-nature/

introductions throughout the world. It has been reported in Argentina, Australia, Tasmania and New Zealand, as well as in Europe (France and Spain). The species was brought to the Thau lagoon in 1971 with the massive introductions of Pacific oysters at that time. It was subsequently deliberately introduced for seaweed farming purposes in Brittany (1983) and Spain. Since then, it has colonised the coasts of Brittany and the south-west Atlantic, as well as the North Sea (1986) and Spain (1988). The cultivation and ability of the species to develop by clinging to artificial and mobile structures (boat hulls, mollusc shells, floating objects) make it one of the most invasive species of algae, colonising a wide variety of environments. It grows on any substrate. Its spread is facilitated by the transfer of oyster stocks. Its spores are able to go dormant at high temperatures (surviving six months in the dark), facilitating its spread. The ecological repercussions vary according to the areas colonised: local biodiversity can be affected by its massive and dense expansion, and by its competition with native species, resulting in a change in ecosystem functionality. It can also facilitate the development of other species by creating a particular habitat (making it a so-called 'engineer' species), particularly because of its size, as it can attain 3 m in length. However, its proliferation leads to the economic costs of cleaning colonised infrastructures and can limit the growth performance of other aquaculture species.

Caulerpa

Biological invasions resulting from the introduction of several species of tropical green algae of the genus *Caulerpa* into the Mediterranean Sea have been widely publicised since the 1980s (Meinesz *et al.*, 2001). Native to tropical waters (Indo-Pacific, Caribbean and African coasts), these algae have been the subject of extensive international trade for aquarium purposes, particularly in the case of *Caulerpa taxifolia*. A clone of this species, accidentally introduced into Monaco waters in 1984, rapidly proliferated to cover 1,300 ha in 1993, 4,630 ha in 1998 and more than 6,000 ha in 2000, spreading to many countries in the Mediterranean basin. *Caulerpa* is tolerant of a wide range of temperatures (7 to 30°C) and its stolon grows by 2 to 3 cm a day, reaching over 350 m/m². Each plant can produce new fronds every two days. New *Caulerpa* meadows have emerged with densities of 5,000 to 14,000 fronds/m² (record 95,000 fronds/m²). With no herbivores to consume it, the alga has spread through vegetative fragmentation, a process further accelerated by human activities such as pleasure boating (anchors) and fishing (nets). The very dense spatial coverage, which tends to homogenise the environment, has caused local extinctions, species displacement, changes in benthic communities and, in particular, to spatial competition with protected Neptune grass (*Posidonia*) habitats of high biodiversity value. The density and size of the Neptune grass also decrease, with chlorosis and necrosis appearing and plants dying. The ecological impact is a massive 55% reduction in algal diversity, with a decline in Neptune grass and a loss of fish populations. The physico-chemical environment has also been thrown out of balance, as have fishing activities, disrupted by the clogging of fishing gear and a reduction in catches, and tourist activities (scuba diving). The seaweed modifies the very structure of the colonised habitats by encouraging the accumulation of sediment and the appearance of algal mats, monopolising space to the detriment of species that develop upwards into the water column. Locally, the effects of the invasion persist even after the algae have been removed, with habitats being restored slowly. Despite its unexplained regression since the 2010s, the environmental impacts remain, with the accidental introduction, probably via ballast water and/or aquaristics, of sea grapes (*Caulerpa racemosa*), which came from south-western Australia in the 1990s and colonise the seabed at depths of up to 70 m (Klein and Verlaque, 2008). Other genetic variants of *C. taxifolia* have also been identified in the Mediterranean, as has another exotic *Caulerpa* species, *C. distichophylla*, with similar impacts (Jongma *et al.*, 2013).

Cordgrasses

The history of cordgrasses (*Spartina*, largely integrated into the genus *Sporobolus* since 2014) in Europe is a model in terms of plant genetics. These are halophytic grasses that live in coastal environments on mudflats (or *slikke*) and the upper part of the foreshore (or *schorre*). Small cordgrass (*Spartina maritima*, now *Sporobolus maritimus*) is native to European coasts. At the beginning of the 19th century (1806–1829), *S. alterniflorus*, a cordgrass originally from North America, was introduced to France and England via the solid ballast of ships. It hybridised with the native species to make a sterile hybrid, Townsend's cordgrass (*S. townsendii*), which spread by vegetative propagation. After 1890, fertile plants appeared, resulting from the doubling of the number of chromosomes in this species to give common cordgrass (*S. anglicus*). At the beginning of the 20th century, these cordgrasses were used to stabilise riverbanks. What appeared to be a nature-based solution in the short term turned out to be a nuisance in the long term, an aspect that, like climate change projections, needs to be taken into account when deciding how to voluntarily introduce a species. *Sporobolous* has become invasive by colonising mudflats because of its high tolerance of variations in salinity (10 to 60 g/l), its high reproductive capacity, with high production of highly fertile pollen, and its ability to multiply by fragmentation. The seeds can enter a dormant phase (< 1 year) before developing to produce dry biomasses of over 1 kg/m². In fact, these cordgrasses can be found from salt marshes to tidal freshwater marshes (Goulletquer, 2016). They profoundly modify the ecosystems they colonise. *Sporobolous anglicus*, in particular has an enlarged ecological niche and a high sediment trapping capacity due to its root system (rhizome) and resistant foliage, causing areas to silt up and replacing local species, including protected saltwort and eelgrass beds (*Zostera* sp.) (Sparfel *et al.*, 2005). Shore and foreshore birds suffer a reduction in feeding grounds as these ecosystems are restructured, leading to more uniform vegetation but also creating new habitats for other species. We therefore see significant impacts at species level (speciation) and on ecological communities, as well as on the functionality of the foreshore and marshes, affecting human uses and activities. The impact of introduced species on the genetic characteristics of native species, such as hybridisation and resulting changes in physiological performance, has been demonstrated not only for plants but also for some fish (salmonids) and shellfish species.

Mnemiopsis

The biological invasion resulting from the proliferation of *Mnemiopsis leidyi*, a jellyfish-like North American ctenophore, is archetypal in terms of its impact on the environment and human activities. The species was accidentally introduced into the Black Sea and Sea of Azov in the early 1980s through the deballasting of merchant ships. With no predators, it proliferated in the rich waters of these seas reaching significant biomass, of the order of one million tonnes in 1989, with densities of over 500 individuals/m^3. Because the species is predatory, consuming fish eggs and larvae, but also on the zooplankton necessary for the development of juvenile fish, fish stocks are collapsing. Significant declines have been observed in both the specific composition and abundance of zooplankton. The trophic chain was profoundly altered and anchovy fishing came to a virtual standstill in 1994. By 1992, annual commercial losses were estimated at more than US$240 million (Pitois and Shiganova, 2015). Secondary introductions into the Caspian Sea have had similar impacts, even affecting seal populations through lack of food. The situation was reversed with the appearance of another ctenophore, *Beroe ovata* — a predator of *Mnemiopsis leidyi* — in the Black Sea in 1997. It is unknown whether this introduction was accidental or deliberate. A voluntary introduction could be likened to biological control, but the uncertainties and risks in the marine environment associated with the introduction of an exotic species mean that such a management option cannot be recommended. *Mnemiopsis leidyi* has continued to expand: it was introduced secondarily via ballast waterways to the Baltic Sea and through multiple additional introductions, it has colonised the coasts of Europe as far as Cherbourg, favoured by a planktonic lifestyle. More recently, populations have been reported on the Atlantic coast. The impact of mnemiopsis on plankton communities along the Atlantic coast and in the Baltic Sea appears to be more inconsistent, limiting the negative effects on populations of species of commercial interest such as cod, herring and sprat (Schaber *et al.*, 2011). In the Mediterranean, as a result of tropicalisation and secondary transfer, the species has reached the French Mediterranean lagoons, where its development is affecting traditional fisheries through predation on local species and the physical clogging of fishing gear.

Common slipper limpet

Native to the east coast of North America, the common slipper limpet (*Crepidula fornicata*) was accidentally introduced to the coasts of the United Kingdom in 1872 through the commercial importation of Eastern oysters. Finding a favourable environment with no predators or disease, it proliferated on European coasts, with secondary introductions facilitated by the trade and transfer of European and Pacific oysters. The internationalisation of shellfish production since the 1980s has also encouraged its expansion. Locally, fishing activities involving dredging contribute to its dispersal through the dumping of sorting waste. More than a century after its introduction, slipper limpet populations are still having a major impact on the nature of the seabed. Its colonies modify the granulometry by trapping particles. It encourages local siltation and its biomasses, which can reach 8 to 10 kg/m^2, lead to a uniformisation of the seabed and hypoxia in the lower layers, with the disappearance of benthic fauna and flora; commercial species such as the warty venus or scallop have thus seen their yields fall sharply. The deterioration of the seabed is irreversible when 50% of the habitat is covered. In the Gulf of Saint Malo alone, slipper limpets represent a biomass of over 200,000 tonnes, limiting any practical management approach. The particularity of this species is that it feeds as both a grazer and a filter feeder. The latter leads it to compete for food with filter-feeding species (mussels, oysters). This food competition, combined with its spatial coverage of deep-water and intertidal shellfish farming concessions and the increased sorting and cleaning of shellfish products necessary on land, affects shellfish farming. In response, this industry invests in dredging and destroying this invasive species. The economic costs involved are significant, both in terms of the management methods required and the loss of production yields. Attempts to exploit the species itself as a product (e.g. limestone amendment, biomaterials, marketing of shellfish in Asia) are still in their infancy.

Pacific oyster

The Pacific oyster (*Magallana gigas*) was voluntarily introduced to France in the 1960s, then massively imported in the early 1970s during the 'Résur' operation to support shellfish farming, which had been devastated by the dying out of the Portuguese oyster (another exotic species). Pacific oyster became invasive in the 1990s as a result of climate change and stock transfers across Europe. Today, it is the most heavily fished and farmed shellfish species in European waters, with annual tonnages of around 150,000 tonnes and an economic value of around €300 million in France. It is also fished recreationally (although sometimes causes injuries from cuts). It should be noted that in oyster farming, the native European flat oyster suffered a collapse in production following the accidental introduction of the exotic parasite *Bonamia ostreae*, the cause of Bonamiosis. Originating in North America, this flat oyster parasite has spread to all European waters since the 1980s through secondary introductions linked to stock transfers.

Although massively introduced into shellfish farming basins along the various coasts of France, only populations of Pacific oysters located to the south of the Loire (Bourgneuf Bay, Pertuis Charentais, Arcachon basin) were able to develop sustainably in the first few years after introduction. The creation of natural beds and the identification/protection of sanctuaries made it possible to maintain the species and support the annual collection of oyster spat necessary for the development of shellfish farming in France after its introduction. With regular annual summer reproduction, a massive reproductive strategy (40–50 M oocytes per spawning individual) and a pelagic larval development phase lasting around twenty days to facilitate dispersal, the population then built up rapidly, supporting the restoration of oyster aquaculture. This introduction has had positive effects on socio-economic development to the present day. However, despite pre-treatment (e.g. drying, brine baths), a number of exotic organisms were accidentally introduced at the same time as Pacific oysters brought from British Columbia (Canada) and Sendai (Japan): their shells also provide a favourable habitat for many species (Gruet *et al.*, 1976). Similarly, several pathogenic species such as the protist

Haplosporidium nelsoni were introduced at the same time, but only detected much later. At the time, the main considerations were the economic aspect of supporting a shellfish industry in crisis, with management methods focusing essentially on human and animal health aspects. The issue of biological invasions and environmental impact of these introductions was little known.

Since the 1990s, the species has gradually colonised European foreshores, developing natural populations as far north as the Norwegian fjords (61° North latitude) as a result of climate change and its effects (Thomas *et al.*, 2016). In addition, the commercial flow of stocks between the various European shellfish-growing regions (introduction route) has facilitated secondary introductions of several dozen species (Wolff and Reise, 2002; Mineur *et al.*, 2014). The oyster shell then serves as a substrate that can be colonized by many species. Using spectral and hyperspectral analysis, Barillé *et al.* (2017) showed that an oyster shell could host more than 90 taxa of photosynthetic endobionts and epibionts.

In the Wadden Sea, the species has supplanted the European flat oyster, replaced the mussel beds initially exploited there and caused a major change in the structure of benthic communities (Kochmann *et al.*, 2008; Markert *et al.*, 2013). The reduced availability of mussels as a food resource for avifauna has not been compensated by the presence of this oyster, whose shells are more resistant to predation, even though adaptive behaviour of gulls has been reported (Cadée, 2001; Waser *et al.*, 2016; Herbert *et al.*, 2018). As an engineer species, its massive colonisation of rocky foreshores tends to profoundly modify local biodiversity and its environment, while promoting recreational fishing. It forms a 'rocky' habitat by building reefs, where previously only benthic communities developed on muddy foreshores. With its filter-feeding mode, its growth modifies the characteristics of both the water column (e.g. reduction in phytoplankton primary production) and sediments (e.g. bio-deposits). Its high filtration capacity makes it a food competitor for other local filter-feeding species. However, these reefs can also be a favourable habitat for numerous species due to the complexity of the reef structures and the heterogeneity of the habitat created.

The Japanese oyster drill

Among the exotic species accidentally introduced to France with Pacific oysters during the shellfish farming recovery plan of the 1970s was the Japanese oyster drill (*Ocinebrellus inornatus*). Native to the Asian Pacific (Korea and South Japan), the species was first introduced to the west coast of the United States during oyster transfers in the 1920s. Identified on the foreshore of the Ile de Ré (France) in 1994, analysis of the genetic characteristics demonstrated that the initial introduction had taken place more than twenty years earlier, associated with Pacific oyster broodstock from British Columbia (Canada) (Pigeot *et al.*, 2000; Martel *et al.*, 2004a; 2004b). This illustrates the high variability in possible lag times shown in **Figure 1**. The effects of climate change in the 1990s are very probably responsible for the emergence of the invasive nature of the species (threshold effect). The species subsequently colonised the various French and European shellfish farming basins through oyster transfers. The species has had a variety of impacts: it uses the same ecological niche as the native sting winkle (*Ocenebra erinacea*) and tends to supplant it because of its physiological characteristics (longer spawning period and reproductive effort, better energy efficiency). Its ecological impact is therefore significant, both in terms of ecosystem functioning and the gradual replacement of the local species. *Ocinebrellus inornatus* is a particularly effective predator of oysters and other shellfish, which it attacks by chemically and mechanically piercing the shell (acid secretion and rasping tongue boring like a drill). In fact, its activity has a direct socio-economic effect both through the loss of shellfish production yields and the costly management resources required. Also of note on the Atlantic coast and in the Mediterranean is the veined rapa whelk (*Rapana venosa*), a gastropod of Asian origin that also predates shellfish.

American blue crab

More recently, the arrival of the American blue crab (*Callinectes sapidus*) on the French Mediterranean coast (Corsica and continental lagoons) has attracted a great deal of media attention because of its impact on both human activities and the environment. This is not the first invasive exotic crab in France, because the Chinese mitten crab has been present there in since the 1930s, and its life cycle is marked by a massive migration phase from freshwater to the marine environment, where the larval phases develop. The species is cyclically invasive and highly tolerant of environmental conditions. Its activity causes bank erosion, digging deep galleries (80 cm) and clogging water pipes (Goulletquer, 2016). As an opportunistic predator, it also disrupts ecosystems and competes with native species.

The native range of the American blue crab extends from the east coast of North America to Argentina (tropical to temperate), where it is the subject of intensive commercial fisheries. The species was reported in Europe from time to time since the beginning of the 20th century, probably as a result of deballasting/biofouling, but without any permanent populations developing. Introductions to the Aegean Sea in the 1930s led to perennial populations, which have since been commercially exploited. The process of biological invasion and demographic explosion of the crab in the Mediterranean dates back to the early 2000s, with strong spatial expansion reaching the Moroccan coast in 2017. To date, more than 17 Mediterranean countries are affected by this invasion, following significant impacts in the north-western basin. In recent years, and following the first isolated reports in Corsica, the American blue crab has extended its range along the continental French coast, particularly along the Gulf of Lion (Labrune *et al.*, 2019). Its presence had already been noted occasionally in the Etang de Berre in 1962 (Galil *et al.*, 2002), but its real expansion on the French Mediterranean coast was not confirmed until 2016. Since then, it has been observed in 15 lagoons in the Occitanie region, three lagoons in the PACA region and 8 lagoons in the Corsica region. It has also been observed at sea and in certain river mouths (Occitanie, Corsica). Being particularly tolerant of environmental conditions, it invades a variety of environments. This crab mates in a low-salinity environment, but releases its larvae in the open sea, where the first stages of development take place, before the young return to the lagoon. An excellent swimmer, the American blue crab is a highly aggressive omnivorous predator, including cannibalism.

It is capable of destroying fishing nets to reach its prey, may weigh up to 1 kg and can travel 15 km a day. Octopuses are considered to be the crab's only predators and are not present in these coastal lagoons. As an invasive species itself, this crab is also a major disrupter of the ecosystem, consuming all kinds of species (shellfish, eels, green crabs, sea bream, sole, mullet, but also amphibians) and destroying fishing gear. In the Etang de Canet in Roussillon, just a few individual blue crabs were reported in 2017, but 10 tonnes were fished in 2020. Fishermen ended up abandoning their traditional activities, while still catching up to 600 kg of blue crab a day. In 2022, these crabs proliferated to an exceptional degree on the east coast of Corsica, with fishing activities in the Etang de Biguglia particularly affected. Mullet fishing was halted, as was eel fishing, and the crabs destroyed fishing nets. This species therefore poses a serious threat to the biodiversity-rich lagoon ecosystems of the entire Mediterranean.

Lionfish

The lionfish *Pterois volitans* and *P. miles* (red lionfish and common lionfish) are voracious generalist predators. *Pterois volitans* is usually found in the Pacific and eastern Indian Oceans at depths of between 2 and 55 metres. In the rest of the Indian Ocean and in the Red Sea, one finds the very morphologically similar *P. miles* (Bottacini *et al.*, 2024), AKA the 'devil firefish'.

Both of these lionfish species have invaded tropical and subtropical areas of the western Atlantic, from the coasts of Brazil to Cape Cod in the United States, with very significant impacts in the Caribbean. The first detection was made in Florida in 1985, probably following several escapes from private and commercial aquariums damaged by hurricanes. Since 2000, the invasion of the Caribbean has been particularly worrying. An unrivalled predator, this species feeds on a wide range of fish, with around a hundred prey identified. Its large stomach capacity enables it to consume substantial quantities, including herbivores that play a key role in maintaining reef ecosystems by controlling algal growth. Up to 90% fish losses, both herbivorous and piscivorous species, have been observed in invaded areas (Ingeman, 2016). In the absence of herbivores, the reef ecosystem is restructured, with algae becoming dominant to the detriment of coral and sponge populations (Kindinger and Albins, 2017). This change in the ecosystem is eliminating traditional refuge areas for many juvenile fish of ecological and commercial interest (DeRoy *et al.*, 2020). With a life expectancy of up to 30 years, lionfish lay more than 2 million eggs a year (a few thousand every 2 to 3 days) from their first year and reach densities five times higher than in their native range (400/ha). As a result, the structure of ecosystems is altered, fishing activities are disrupted — economic losses of around €10m/year have been estimated in the West Indies — and the species can cause harm to humans when caught due to its venomous spines.

In the Mediterranean, *P. miles* arrived from the Red Sea and the Indian Ocean via the Suez Canal and/or the ballast tanks of ships. After first being reported in Israel in 1991, its progress has accelerated since 2011, gradually invading the Mediterranean as this sea becomes more tropical. Recent sightings have identified it as far west as Cyprus, with a westward progression predicted as climatic conditions become more favourable in the future. With no predators, a higher reproductive capacity than the species in its native range and an opportunistic diet, it has a particularly significant impact on both ecosystems and fishing activities.

IMPORTANCE OF THE ECOSYSTEM APPROACH AND ANALYSES

These different examples highlight the heterogeneity of impacts, which explains the difficulties with the management methods required. The need for scientific information on the biological traits of exotic species is essential here, and not just on a regional scale. It is now crucial to conduct comparative analyses of the species' genetic characteristics using samples from all known populations worldwide. These would provide information on the pathways, vectors and temporal dynamics of introductions, but also facilitate management decisions. Biological characteristics and traits also provide a better understanding of the dynamics and spatio-temporal trajectory of invasions, helping in particular to identify threshold effects in response to the effects of climate change. All these factors increase the relevance of risk assessments associated with these exotic species.

We have seen that the impacts of biological invasions are measured at the level of populations and ecosystems, which fully justifies the development of scientific studies on the specific interactions and functioning of ecosystems using an 'ecosystem approach'. These studies need to be conducted over the long term, as Strayer *et al.* (2006) emphasised. Following an analysis of almost 200 cases of biological invasions, they found that very few studies assessed the impacts on a repeated basis, and 40% of these did not take into account the duration of the invasion, with the majority of studies being of short duration. They concluded that the frequency of monitoring and its brevity (duration of each monitoring) made it insufficient to accurately reveal the effects and impacts of these invasions.

Furthermore, the impacts of these biological invasions can be both 'positive' and 'negative', requiring an integrated analysis approach as recommended by the European MAES project (2020) and the French Efese assessment: with the aim of maintaining ecosystem services, it is also necessary to assess the sustainability thresholds of the 'bouquets of services', those that interact with each other, but also to take account of both 'market' and 'non-market' services in order to avoid maximising the value of a service to the detriment of biodiversity and/or its resilience. This implies the participation of all stakeholders in the impact assessment process, using scientific approaches developed in the human and social sciences.

HOW CAN WE MANAGE BIOLOGICAL INVASIONS?

Experience has shown that managing exotic species in the marine environment poses far greater challenges than those encountered on land. The difficulties are due to the continuity of the marine environment, which requires an 'ecoregional' approach to the problem, going beyond regional and national contingencies. In addition, any treatment can extend beyond simple administrative boundaries. In addition, access to the marine environment is fairly limited and requires considerable logistical resources. It is therefore imperative to consider *prevention* as the priority approach, far preferable and the most economical, in order to prevent invasions before they take place, rather than deploying control methods aimed at eradicating populations and species that are already established. Unfortunately, in the majority of cases, the decisions taken by managers are late in coming, and arrive when negative effects are already beginning to weigh on the environment and human activities. However, it has been clearly demonstrated that impact-based management alone is problematic and that a precautionary approach is needed, with not only anticipatory management methods at species entry points, but also consideration of prioritising future impacts (Ojaveer *et al.*, 2015). The absence of exhaustive scientific data and the associated uncertainties do not justify the absence of decision-making. For example, a single sighting of an exotic species does not necessarily mean that there are many individuals or even permanent populations, meaning that further observation is required. It does, however, allow a rapid response. According to this precautionary approach, the management procedures actually cover any arrival of exotic species, whether or not they are subsequently invasive. This highlights the importance for scientists and managers of having access to international databases that provide information on the intrinsic characteristics of the species concerned, particularly their potential invasiveness!

When it comes to managing exotic species, it is also necessary to differentiate between voluntary and accidental introductions.

VOLUNTARY INTRODUCTIONS

The first cases were often observed during the aquaculture developments of the 1980s. A large number of fish, mollusc and shrimp species were tested in controlled and/or open environments as a source of economic revenue, but this subsequently led to feral populations and even to the introduction of associated parasites or pathogens. Indeed, all aquaculture farming is a source of escapes, with the exception of aquaculture practices in strictly closed environments or quarantine. Despite the adoption of initial sanitary and animal health measures, many exotic species, with their parasites and pathogens, have been introduced simultaneously. These aquaculture activities have also been the source of secondary introductions through the movement of livestock. Most international conventions, such as the CBD and the FAO Code of Conduct for Responsible Fisheries, recommend the use of native species for aquaculture purposes. The impact of the introduction of exotic pathogens on aquaculture and fisheries has also led to the setting up of international working groups to deal with this issue: these include the Ospar Convention (Oslo-Paris) and ICES for the North Atlantic, with the Working Group on the Introduction and Transfer of Marine Organisms (WG-ITMO). These groups of experts contributed to the drafting of European regulations such as the EU Council Regulation 708/2007 of 11 June 2007 concerning use of alien and locally absent species in aquaculture.[12] It aims to regulate the diversification of species farmed in aquaculture, while remaining vigilant with regard to the introduction of species that could prove harmful to ecosystems, and this within the framework of the European single market in order to avoid any distortion of competition between countries. A restrictive

---

12. https://eur-lex.europa.eu/legal-content/FR/TXT/PDF/?uri=CELEX:32008R0535&from=FR

technical framework governs aquaculture practices with regard to the introduction of exotic species and locally absent species (animals, plants or microorganisms, but also fertile polyploids such as tetraploid oysters). It requires project developers to follow procedures for analysing risks and developing measures based on the principles of prevention and precaution. In addition, emergency plans should be drawn up in case of need. All of this must be specified in applications for authorisation from the competent authority. Any such project is therefore subject to the issue of an introduction permit by the Member State of destination. It should be noted that neighbouring European countries, which could potentially be affected by the measure, are entitled to give their opinion. Member States may also enact stricter regulations, the European text being the 'lowest common denominator'. However, there are two exceptions to this strict regulatory framework, the details of which are also set out in the French Environment Code:

- The first exception is the species listed in Appendix IV of the EU regulation, which correspond to taxa already widely exploited in Europe in 2007, such as the Pacific oyster, Manila clam or, in freshwater, the common carp (*Cyprinus carpio*), wels catfish (*Silurus glanis*) or giant river prawn (*Macrobrachium rosenbergii*) in overseas waters, despite their invasive nature. It seemed difficult to adopt a retroactive regulatory approach for these exploited species, which are considered to be already 'naturalised'. However, it should be emphasised that new cultures of species listed in Appendix IV in open areas may increase the pressure on the environment and local biodiversity. Furthermore, a decision to implement restrictions on the aquaculture of these species, as currently envisaged by the United Kingdom for the farming of Pacific oysters, would be ineffective, given that there are already wild populations with high reproductive capacity present beyond the country, on a regional and ecoregional scale (Shakspeare *et al.*, 2024);
- The second exception is closed and secure aquaculture facilities, where the risk of escape is virtually nil. In this case, except in special circumstances, no prior environmental risk

assessment is required, and no permit is issued. However, the facility must be registered with the prefecture on the list of closed aquaculture facilities. These closed facilities generally use recirculating water systems, with all equipment housed in fully secure above-ground tanks that prevent any spillage into the environment. Since 2007, aquaculture developments in France involving exotic species, particularly farmed shrimp and algae, have primarily operated under this model. However, it remains debatable whether recent aquaculture installations (since 2017), used for farming exotic species (e.g. shrimp) in freshwater and brackish marshes, are truly 'closed'.

It should also be noted that there are generic procedures in France aimed at limiting environmental impacts: classified installations for environmental protection (ICPE procedures).[13] More specifically, the technical arrangements for holding fertile polyploid oysters (*M. gigas* 4N) are also governed by the Order of the Ministry of Agriculture and Food of 7 December 2021,[14] a regulatory text that requires holders to certify their facilities where fertile polyploid oysters must be confined.

In addition, and still in connection with aquaculture, the WOAH draws up codes and manuals specifying standards for improving the health and welfare of animals and veterinary public health worldwide, including standards for international trade in animals and their products. Certain exotic parasites and pathogens, responsible for known or emerging diseases, are subject to compulsory declaration by member countries, which must implement procedures to limit the transfer and establishment of these pathogenic species. This is the case for the exotic protist parasite *Bonamia ostreae*, which has drastically affected native European flat oyster populations since the 1980s, with the classification of areas to limit transfers of infected oysters to areas free of this parasite.

---

13. https://entreprendre.service-public.fr/vosdroits/F33414
14. https://www.legifrance.gouv.fr/download/pdf?id=cl6xClNROSVDIYXS3-4umQTtqKeN8XbR9fRT6gSr4I4

These various measures lead us to believe that the voluntary introduction of new species is relatively well controlled, limiting the environmental and economic risks for the future. However, there are two additional aspects to consider:

- Aquaculture species may be the subject of genetic selection programmes likely to weaken wild populations at a later date in the event of escapes from aquaculture facilities. This calls for continued vigilance, as illustrated by Perriman *et al.* (2022) with regard to salmon hybrids from crosses between selected strains and wild populations having reduced fitness. Salmon farming in Canada is in trouble because of its impact on native salmon species, which are currently critically endangered;
- Climate change and the warming of marine waters are new factors when we consider the decisions taken in the past concerning cultivated species that could not reproduce in their initial environmental context. These environmental trajectories, which were not taken into account at the time, should lead managers to reassess these situations and strengthen the regulatory criteria. This justifies, for example, taking climate projection models into account in risk assessments for introductions, as proposed in the AS-ISK procedures (Copp *et al.*, 2016; Vilizzi *et al.*, 2021).

UNINTENTIONAL OR ACCIDENTAL INTRODUCTIONS

Unlike voluntary introductions, unintentional introductions are especially concerning and must be made a priority to strengthen management measures and develop new approaches. Although the European regulatory framework is very explicit in the 2014 text, covering the problem from prevention to the restoration of affected ecosystems, its implementation is still only very partial.

As a reminder, **Figure 1** shows the management methods used at the different stages from introduction to the invasion process, along a continuum: prevention, checks at points of entry, rapid response, containment and control/regulation.

In the case of accidental introductions, prevention is the most viable management option to avoid the undesirable effects of these alien species (Galil *et al.*, 2019). Once a species has become established, management options in the marine environment are relatively ineffective and generally costly (Booy *et al.*, 2020; Lehtiniemi *et al.*, 2015). Prevention requires the implementation of various complementary approaches. Scientific knowledge on high-risk alien species, combined with risk analyses and international databases (e.g. GISD, EASIN, AquaNIS[15]) provides a foundation for stakeholder dialogue and the development of practical communication tools (e.g. information sheets) (Olenin *et al.*, 2014). The resource centre co-piloted by the French Office for Biodiversity (OFB) and French committee of the IUCN (Centre de ressources — Espèces exotiques envahissantes [CDR-EEE][16]), launched in 2018, serves as a collaborative platform and mediation tool to support stakeholders in addressing the challenge of biological invasions.

PREVENTION AND REGULATORY MEASURES

Regulatory measures complement this awareness-raising by implementing practices and techniques to deal with the issue upstream. At international level, the IMO approved the Ballast Water Management (BWM) Convention in 2004. This made the preventive treatment of ballast water mandatory, requiring more than 50,000 ships to attain given technical standards. The convention required the signature of 30 countries, representing at least 35% of world tonnage. It has been operational since 2017,[17] and was fully implemented in September 2024. The convention has been (and remains) the source of numerous technological innovations to facilitate these treatments, such as the implementation of specific ultraviolet treatments.[18] This development on an international scale is certainly the most significant improvement in marine environmental protection made

---

15. https://www.iucngisd.org/gisd/; https://easin.jrc.ec.europa.eu/easin; https://aquanisresearch.com/

16. https://especes-exotiques-envahissantes.fr/

17. https://www.imo.org/fr/MediaCentre/HotTopics/Pages/Implementing-the-BWM-Convention.aspx

18. https://www.ballast-water-treatment.com

in recent decades, even if there is still room for improvement. However, the lack of treatment for marine biofouling remains a problem. Devices that use copper, aluminium and iron anodes to induce an electric current to reduce biofouling unfortunately only provide a partial solution, and are also a source of pollutants. Several countries have adopted even more restrictive approaches, such as Australia, where a zero-tolerance policy has been implemented in terms of biosafety. On Barrow Island, a risk analysis system is in place for all equipment and personnel arriving on the island, and boat hulls are inspected. A quarantine policy has also been adopted to prevent any introduction of exotic species on land and in marine ecosystems. In 2023, Australia banned a cruise ship from docking for six days because its hull had not had antifouling treatment; the same ship had previously been sanctioned in New Zealand for the same reason. This vessel had to anchor 17 miles off the coast and have its hull cleaned by a team of divers. Such approaches are particularly well suited to the conservation of protected areas in sanctuaries and/or zones of high protection.

While these developments have undoubtedly led to progress, there are still significant shortcomings in terms of prevention: there are no operational technical measures in place to guard against the introduction of species into major waterways such as the Suez and Panama Canals. However, international databases on invasive alien species (GISD, EASIN, AquaNIS), information systems and maritime traffic analysis provide information prior to any arrivals in ports, making it easier to assess the risks associated with these species and develop priorities for action. For example, knowledge of the current distribution of invasive exotic species at the European level via EASIN is helping the competent authorities adapt their surveillance systems to meet regulatory requirements for rapid detection plans (Magliozzi *et al.*, 2023) and helping managers to prioritise risk assessment areas and the actions to be taken (Magliozzi *et al.*, 2024).

Spurred on by the recommendations of international conventions and the findings of scientists, such as the CBD (decision XII/16) (Nagoya Agreements, 2010), the Bonn and Bern Conventions, CITES, FAO, UNESCO and the IUCN assessments, the regulatory context has been strengthened at European and French levels. In 2008, the European Community defined its strategy for restoring the marine environment through the MSFD. It aims to achieve good environmental status across the board, with one of the descriptors focusing on the control of non-native species (Olenin *et al.*, 2009). More recently, the 2014 regulation is devoted exclusively to this issue. In particular, it establishes a list of exotic species of Community concern and requires risk analyses to be carried out for them.[19] This list currently includes 88 species, of which only four are marine, the brown alga *Rugulopteryx okamurae*, black pygmy mussel (*Xenostrobus securis*), Chinese mitten crab (which has a marine development phase) and striped eel catfish (*Plotosus lineatus*), and is supplemented by national lists. France first adapted its legislation with the 2016 Law on the Restoration of Biodiversity, and then developed a national strategy on IAS, the first pillar of which focuses specifically on prevention,[20] and formalised an action plan (2022–2030) comprising 19 items.[21] The Environment Code has been updated with two articles: art. L. 411-5, level 1 "Prohibition on the introduction into the natural environment of non-domesticated, non-cultivated animals and plants not indigenous to the territory" and art. L. 411-6, level 2 'cumulative bans', which is more restrictive, and by *ad hoc* lists of species drawn up for mainland France and French overseas territories. These lists have been updated since 2018 and include the American blue crab (level 1) and African blue swimming crab (*Portunus segnis*) (level 2).[22] This applies *de facto* to species traded internationally (e.g. for aquariums). It should be noted that the National Biodiversity Strategy (SNB3, 2023) includes a measure (No. 10) in its Axis 1 on limiting the introduction of new IAS through a preventive approach. These different regulations should help to achieve the objectives set

---

19. https://eur-lex.europa.eu/legal-content/FR/TXT/PDF/?uri=CELEX:32022R1203
20. https://especes-exotiques-envahissantes.fr/strategie-nationale-relative-aux-eee/
21. https://www.ecologie.gouv.fr/sites/default/files/20220315_EEE_VDEF.pdf
22. http://especes-exotiques-envahissantes.fr/29-especes-exotiques-envahissantes-nouvellement-reglementees/

at the last COP 15 of the Global Biodiversity Framework in Kunming, Montreal (2022), to reduce new introductions by 50% by 2030.

SURVEILLANCE OF ENTRY ROUTES

Despite these preventive approaches, exotic species are still arriving on French coasts via various vectors and routes of introduction, requiring checks at priority points of entry. Seaports (yachting, commercial, military), shellfish farming areas and seafood shipping sites are candidate areas. For many overseas territories, ports are the main import routes for consumer products and are therefore gateways for these alien marine species. Their management and development plans play an essential role. For the marine environment, this introduction phase, prior to the establishment of the exotic species, is the ultimate operational phase enabling these exotic species to be checked for in an inexpensive manner.

The effectiveness of these checks at priority entry points depends on the implementation of a monitoring programme with an appropriate sampling strategy in order to take rapid response measures following detections. This also includes optimised monitoring tools, some of which are still the subject of research projects (e.g. environmental DNA). Monitoring these introductions and their impacts is, however, a prerequisite for the implementation and subsequent choice of management methods. Surveillance and rapid response systems have proved their worth in countries such as the United States, where several species of Asian origin were eradicated as soon as they were detected in West Coast commercial ports. In France, some recent efforts in this direction are noteworthy. As part of cycle 2 (2020–2026) of the MSFD, a sub-programme is dedicated to monitoring French ports and shellfish farming areas, including areas sensitive to biopollution and sites with marine renewable energy infrastructures. Although this is not a real sampling strategy covering the whole of mainland France and all the sites 'at risk', it is nevertheless a monitoring system that is under construction and needs to be expanded. On the other hand, there is no dedicated monitoring system in overseas France. Biodiversity observatories are a potential source of contributions and data. Information from decentralised government departments (Regional Department of the Environment, Planning and Housing, DEAL), management structures such as marine nature parks, as well as the activities of environmental associations, including diving clubs and the growth of participatory science approaches, can partly supplement these observations. It should be noted that these active surveillance, early detection and rapid response approaches are set out in the 2014 regulation as well as in the National Strategy on Invasive Alien Species and the SNB3 (2023) (Axis 1, Measure 10), cited above.

RAISING PUBLIC AWARENESS OF BIOLOGICAL INVASIONS

Raising public awareness, as well as among stakeholders and users of the sea, about the risks associated with the introduction of exotic species is a management priority. Disseminating information to the general public is essential (e.g. scientific communication, mass media). Recreational boaters need to be made aware of the risks of introduction via anchors and unmaintained boat hulls. They therefore need good practices that will slow down or even prevent the spread of exotic species.[23] The same applies to recreational fishing. Similarly, owners of marine aquariums must not release commercial exotic species into the wild. Particular attention must also be paid to professionals in the maritime world (livestock transfers, fishing activities). Raising awareness requires the provision of summary information on the risks associated with these practices, as well as on the best practices to adopt, and the inclusion of species pre-identified as posing a risk. There is an urgent need to strengthen this co-constructed communication approach between stakeholders in order to improve the reception of messages (Courchamp *et al.*, 2017).

---

23. Good careening practice: https://www.eaufrance.fr/sites/default/files/2018-07/Guide-sur-les-bonnes-pratiques-de-carenage-AFB-2017.pdf

Participation in participatory science schemes is also important, as demonstrated by the implementation of observation networks. These include the 'Alien networks', some of which are run by the *Fédération Française d'Études et de Sports Sous-Marins* (FFESSM) and others by the *Base pour l'Inventaire des Observations subaquatiques* (BiObs),[24] which cover several regions including Occitanie, Corsica and Normandy, with the '*Sentinelles de la mer*' initiative, and the *Grand Ouest* (western France). As well as acquiring new knowledge, their aim is to strengthen and structure prevention, raise awareness and set up rapid reaction systems backed up by operational surveillance. One project is specifically dedicated to the observation of IAS on pleasure craft in the Occitanie region. The OFB is launching an initiative to coordinate these networks in order to capitalise on these particularly important sources of information. All these data from different sources contribute to the information base needed to make decisions and implement action and rapid response plans.

ONCE AN INVASIVE SPECIES IS ESTABLISHED WHAT MEANS CAN BE USED FOR ITS CONTROL?

Once the species has become established in the marine environment, the remaining management options involve controlling invasive populations, generally through recurrent, long-term 'trimming back' of their numbers. Theoretically, there are three possible options for achieving this objective: biological control, direct destruction of populations or, ultimately, their exploitation. However, no case of biological control has been documented in the marine environment (IPBES, 2023). This approach was once envisaged to combat the proliferation of *Caulerpa* in the Mediterranean, notably by introducing another exotic species native to the alga's area of origin. Two ascoglossan molluscs that consume exclusively this alga, *Oxynoe azuropunctata* and *Elysia subornata*, had been identified as candidate species and *E. subornata* was pinpointed as the largest consumer of *Caulerpa*

---

24. https://biologie.ffessm.fr/reseaux-alien, https://bioobs.fr/blog/

taxifolia. The introduction of an additional exotic species into an area considered to be a biodiversity 'hot spot', and whose future was not predictable, was considered as a high risk by the Academy of Sciences, which rejected this management option. Subsequent developments have shown that this decision was justified. Our current scientific understanding of marine biological invasions only reinforces the rejection of this type of option.

Destroying invasive populations

The destruction of populations by different methods, particularly physical ones such as uprooting, trapping, capturing, or by bounty and reward programmes, is still the most common, but also the most expensive. This last approach involves appealing to the public to reduce populations, offering rewards that may sometimes be financial:

- Japanese oyster drill fishing was encouraged in the Marennes-Oléron basin in the 1990s;
- Lionfish sport fishing competitions have been organised in the Caribbean;
- Between 1999 and 2009, fishing for veined rapa whelk in Chesapeake Bay (USA) was rewarded with a bounty of US$2 per shell and US$5 per live animal, enabling the destruction of 18,000 individuals at an annual cost of US$35,000;
- Green crab fishing in the United States has been funded since 2013 at a rate of US$0.4 per pound (> 2 t caught);
- However, these systems are limited in scope and only relatively effective (Olden, 2024).

Annual operations to destroy populations are generally more massive in scale, with the main aim of maintaining the functionality of the ecosystems concerned and reducing the economic consequences. These operations must also be carried out with ethical considerations in mind. For example, a guide provides managers with a choice of lethal and non-lethal methods for the eradication, management and containment of exotic vertebrates that take animal welfare into account (Smith *et al.*, 2022). In Norway, dams on rivers during the migration phase allow exotic pink salmon to be directed towards sorting devices with image analysis, followed by manual sorting and slaughter in order to reduce their populations (ICAIS, 2024). Among the examples already cited, the case of the common slipper limpet is also of interest: populations of this species have been the subject of annual dredging operations in the Pertuis Charentais (Marennes-Oléron basin) since the 1980s to limit trophic competition with oyster and mussel farms, and to keep shellfish concessions operational. These maintenance operations have kept these populations under control, unlike those in the Bay of Mont-Saint-Michel/Gulf of Saint-Malo, where the lack of management has allowed the species to proliferate to such an extent that any attempt to control it would require a large-scale industrial approach.

Similarly, the proliferation of Pacific oysters in shellfish farming areas such as the Marennes-Oléron basin, Arcachon basin and the Bay of Bourgneuf requires maintenance campaigns, particularly on abandoned shellfish farms, using appropriate technical equipment (dedicated boats, marine-adapted mobile equipment, etc.). Initially organised on an area-by-area basis, under the impetus of professional shellfish farming organisations and in partnership with local authorities, these campaigns are now carried out from the perspective of maintaining and restoring the environment as a whole, with broader implications (e.g. marine park management councils). Also of note on France's Atlantic and Mediterranean coasts is the Australian tubeworm *Ficopomatus enigmaticus*, a species that colonises any hard substrate and forms tubes that become a calcareous mass, rapidly altering the ecosystem physically, chemically and biologically. These concretions can crust ships' hulls and engines, reduce the width of canals and storage water basins, clog pipes and even block locks by colonising port structures. This proliferation leads to significant maintenance costs.

In addition, various invasive exotic algae require removal campaigns, both in the Mediterranean and in the Atlantic, in order to avoid obstructing navigation and to maintain ecosystem functionality (e.g. Japanese wireweed). In protected areas with a high biodiversity heritage value, such as the Port-Cros National Park, campaigns to remove *Caulerpa taxifolia* have enabled this remarkable ecosystem to be preserved in its current state. There is no shortage of management methods for controlling invasive populations because, in most cases, the decisions taken result in significant impacts on the environment and human activities, which transform into economic costs that force the manager to act. Ultimately, these management methods are the most costly, requiring recurrent action over time.

Actions to control populations should be followed by operations to restore affected ecosystems, as defined by the 2014 regulation (art. 20): "Proportionate restoration measures should be implemented to strengthen the resilience of ecosystems to invasions, repair the damage caused and enhance the conservation status of species and their habitats in accordance with Directives 92/43/EEC and 2009/147/EC, the ecological status of inland surface waters, transitional waters, coastal waters and groundwater in accordance with Directive 2000/60/EC and the ecological status of marine waters in accordance with the MSFD. The costs of these restoration measures should be recovered in accordance with the polluter pays principle". Member States must therefore take "appropriate restoration measures in order to contribute to the recovery of [degraded] ecosystems", unless "a cost–benefit analysis demonstrates, on the basis of available data and with a reasonable degree of certainty, that in the long term the costs will be exceptionally high and disproportionate to the benefits of eradication". These measures are particularly justified in natural areas under specific protection (marine protected areas, marine parks, highly protected zones, etc.) and with a high heritage value. However, these restoration operations rarely restore ecosystem conditions to their original state, and may sometimes lead to unforeseen consequences (Hacker and Dethier, 2009)

Creating value with invasive species

An alternative to simple eradication is to exploit invasive species, though this idea continues to divide opinion (French IUCN Committee, 2018). The 2014 European regulation requires an environmental assessment adapted to the risks of dispersal into natural environments, particularly in the case of implementing projects for the commercial exploitation of IAS, even if only species listed at the European level are concerned. However, this management option of commercial exploitation raises ethical and societal questions, as it can also have a beneficial effect as a service provided by nature (e.g. fisheries) (Tsirintanis *et al.*, 2022). As a matter of principle, the development of an economic sector using IAS would need to guarantee the sustainability of the supply of the resource. This point is at odds with management aiming to limit impacts by reducing and controlling populations in order to preserve the environment.

However, different scenarios are possible: co-products may be derived from these proliferations. Active ingredients can be extracted from the biomass produced for biotechnology purposes. For example, hyper-absorbent substances and collagen are extracted from invasive jellyfish in the Mediterranean and macroalgae are being transformed into bioplastics. Trials are also being carried out to recycle Japanese wireweed in Normandy after it has been collected (SMEL, 2016). The development of natural populations of Pacific oysters, as well as farmed products, are a source of biomaterials that can be recycled industrially for a variety of applications (e.g. ceramics, paint, soil improvers, shellfish concrete). These are just a few examples of the diversity of potential applications.

In trying to destroy IAS populations, managers can be faced with very large biomasses of products with no direct pre-established use, but with significant treatment costs. This is the case with the destruction of Chinese mitten crabs in Great Britain, where this species buries itself by digging holes in banks and dykes built to protect against flooding, thus causing major damage and significant economic losses. Belgium has developed a specific trap on an industrial scale, based on the crab's limited swimming ability. The deployment of three traps in Great Britain has resulted in the capture of several million crabs that cannot be commercially exploited for consumption. The biomass recovered is in principle considered as agricultural bio-waste, with the financial costs of processing imposed. Managers are now moving towards using them as bait for common whelk (*Buccinum undatum*) fisheries. In Belgium, similar quantities of Chinese mitten crabs are supplied to zoos to serve as animal feed. In Canada, the invasive European green crab is used as bait for the American lobster (*Homarus americanus*) fishery. However, this latter operation poses risks related to the transmission of pathogens carried by the crabs (100 symbionts identified, including 10 pathogens such as *Aerococcus viridans* var. *homari*) (Gomez-Chiarri and Cobb, 2012; Fraser-Clark, 2024). The Chinese mitten crab has likewise spread the microsporidian parasite *Hepatospora eriocheir* across Europe (Normant-Saremba, 2024). These factors demonstrate the need for precautions in the management of bio-waste through the recovery of these species, and the need for adequate scientific information to support decision-making.

Biological invasions also provide sources of food production: following its introduction to France for aquaculture purposes in the 1980s, the Manila clam developed feral populations that are now fished commercially, with a production of several thousand tonnes a year. In this particular case, the impact of the species remains limited, as it has not systematically supplanted the European species, the grooved carpet shell (*Ruditapes decussatus*), although hybrids do exist (genetic impact).

With no means of controlling lionfish populations in the Caribbean and given the economic losses estimated at €10m per year, managers and environmental associations have turned to fishing for human consumption. Sport fishing competitions have been organised, and restaurant owners have been trained in how to prepare the fish, which nevertheless present a risk because of their venomous spines. Several hundred tonnes of fish have been caught in this way, and an export chain to the American market has developed from the Caribbean. While populations have declined in some areas, such as in the French West Indies, due to overfishing, the creation of a market in the United States has stimulated demand, which now outstrips supply. As a direct consequence an import–export approach has been implemented, with regulation to maintain supplies of lionfish over the long term.

In the Mediterranean Sea, several IAS, including fish, crustaceans and molluscs, are currently being exploited and marketed to a significant degree (Tsirintanis *et al.*, 2022). These include Randall's threadfin bream (*Nemipterus randalli*), narrow-barred Spanish mackerel (*Scomberomorus commerson*), yellowstripe barracuda (*Sphyraena chrysotaenia*), Lessepsian lizardfish (*Saurida lessepsianus*), goatfishes (*Upeneus* spp.), rabbitfishes (*Siganus* spp.)

and crustaceans such as penaeid shrimps (*Penaeus pulchricaudatus* and *Metapenaeus* spp.). Reported for the first time in Antalya Bay (Turkey) in 2013, the northern brown shrimp (*P. aztecus*), which originates from the western Atlantic, has easily colonised the eastern Mediterranean and is fished by bottom trawlers in Egypt, Italy and Sicily.

The African blue swimming crab and American blue crab are case studies in the Mediterranean. The invasion of the Tunisian coast by the former, nicknamed 'daesch' and originating from the Red Sea, caused the collapse of traditional fisheries, which in response developed an export industry of this species, including a processing plant and export to Asia. The second, considered to be one of the 100 worst invasive species, has colonised the Mediterranean as a result of climate change and human activity (17 countries affected). Responses differ from country to country. The species is exploited on a permanent basis in the Adriatic Sea (Italy) but remained relatively localised until the 2000s. More recently, the species has expanded rapidly, with major impacts, including in the Adriatic, where it consumes mussels and clams, including the Manila clam, an exotic species used in aquaculture that has developed feral populations in the Venice lagoon since the 1980s. The crab appeared in the Ebro delta (southern Catalonia) in 2012 and has decimated the species exploited locally. A commercial fishing industry has developed based on the invasive species, with 2.2 million crabs estimated to be caught between 2016 and 2019. Spain exports this crab to China and South Korea, but its market value has collapsed, dropping from €12/kg in 2002 to €8/kg in 2016 and continuing to decline since then. Exploiting such invasive species seems logical at first: increasing the fishing effort will reduce crab populations. However, sustainable fishing strategies are now emerging. For example, fishers releasing reproducing females into favourable environments. Fishing gear needs to be improved, and its structure strengthened to target the species. Canning factories have been set up, contributing to local employment. These short-term socio-economic benefits have pushed ecological issues into the background.

In France, the American blue crab attracts particular attention due to its impact, especially on traditional fisheries in Mediterranean lagoons. From a regulatory point of view, the species is listed at level 1 in article L. 411-5 of the French Environment Code, leaving open the possibility of its commercialisation. It is starting to be used by chefs, particularly in Corsica. At this stage, the main focus is on short supply-chain fishing without maintaining the resource. However, the approach adopted to date is still mainly geared towards providing financial support to fishermen to reimburse their losses, with the prospect of destroying crab populations. Ultimately, the result is that blue crab management methods vary across the Mediterranean, with divergent strategies, whereas an 'ecoregional' and cross-border approach would be necessary to really manage this species, whose individuals, it should be remembered, can travel 15 km a day. This heterogeneity in practices can only limit the effectiveness of a efforts to reduce populations for environmental restoration. The problem is compounded by the fact that, at the request of Algeria, Tunisia and the EU, the General Fisheries Commission for the Mediterranean (GFCM), whose remit covers the Mediterranean, Black Sea and adjacent waters, recently recommended maintaining blue crab fisheries at the maximum sustainable yield (MSY) to guarantee their long-term survival.[25]

So, in the absence of success in controlling their populations, there is a tendency to make use of IAS. However, this type of management requires careful prior thought. The French Committee of the IUCN (2018) proposed such a framework to assess the value of recovering IAS before any action is taken. This is based on four questions, relating to knowledge of the species targeted by the project, the objectives of any project and its integration into an overall management strategy, the identification, anticipation of risks and ecological assessment of the project, and finally the involvement of multiple stakeholders and partners. This non-quantitative approach is a form of self-assessment of the value of projects to develop the profitable use of alien species.

---

25. The FAO defines MSY as the greatest quantity of biomass that can, on average, be continuously extracted from a fish stock under existing environmental conditions, without affecting the reproductive process.

HOW CAN MARINE INVASIVE SPECIES BE SUCCESSFULLY ERADICATED?

As a reminder, eradication means the total and permanent elimination of a population of an IAS by lethal or non-lethal means, as defined by the 2014 EU regulation. Although the voluntary eradication of an invasive species generally remains a real challenge, as in the case of primrose willow in aquatic environments, terrestrial and freshwater ecosystems offer slightly more opportunities than those in marine environments. For example, the waters of Lake Saint-Esprit — the Fréjus reservoir — were invaded in 2018 by the apple snail (*Pomacea* sp.), a South American species known for having previously ravaged the Ebro delta (Spain). Three years after draining the 60,000 m^3 reservoir at this site, the species was considered eradicated, allowing the environment to return to its original state. In fact, in both terrestrial and aquatic environments, 86% of the 1,000 eradication attempts listed worldwide have been successful, including for species that have been established for many years, eradication being a less costly option than long-term management (Simberloff *et al.*, 2013). Despite this work, several IAS remain a major and lasting problem in these ecosystems in France (e.g. Japanese knotweed, *Reynoutria japonica*).

The situation is very different in the marine realm, whose characteristics (very open ecosystems, fluid environment, ocean currents) and whose species (population dynamics, reproduction methods, pelagic larval diffusion, etc.) are all criteria that facilitate the development of populations over large geographical distribution areas, particularly in temperate environments, thus limiting the possibilities for eradication. To date, no complete global extinction of a species has ever been documented in the marine environment once the process of biological invasion is underway (IPBES, 2023).

SOME CASES OF ERADICATION

Successful eradication of marine invasive species is extremely rare worldwide, largely because the openness and fluidity of the marine environment, combined with the reproductive and dispersal strategies of the species concerned, make control particularly difficult. It remains extremely hard to control a species once it has become established, and it is still rare to detect a new species before its population has become established. While eradication may be less costly than controlling proliferation, it can only be successfully done during the initial phase of introduction, when populations cover limited areas and/or when proliferation sites are more or less closed (e.g. marinas). Despite numerous attempts, there have only been around ten successful cases of eradication since the 1950s. One of the first documented, which had a fairly simple approach, was the manual extraction of the whelk *Reishia clavigera* and its oothecae, associated with a Pacific oyster farm in British Columbia in 1951 (Canada) (Carlton, 1979). More recently, temperature treatment was used to control the establishment of wakame, whose sporophytes were present on the hull of a trawler wrecked in the Chatham Islands, New Zealand. However, fifteen months of heat treatment were required at an estimated cost of NZ$0.4m, a very unusual case (Wotton *et al.*, 2004). Since then, eradication strategies have made significant progress with more extensive spatial approaches. Three case studies illustrate the conditions required for successful eradication.

A marine polychaete worm in California, United States

The first implementation of a structured eradication programme was linked to the aquaculture of abalone originating from South Africa and introduced into California (USA) in 1993. A Sabellidae marine polychaete worm (*Terebrasabella heterouncinata*) was accidentally imported with the batches of abalone. As well as affecting the calcification of the farmed shellfish, the worm was also found to infest other species of gastropods in the region, causing shell malformations (Culver and Kuris, 2000). A rapid response was needed to deal with the problem. Considering the epidemiological theory of density-dependency thresholds for effective transmission, a host population reduction strategy was set up, targeting the removal of local populations of the black turban periwinkle (*Tegula funebralis*). The aim was to raise the transmission threshold to such a high level that the recruitment of new individuals would become very limited, if not impossible, due to the disappearance of the host population. More than 1.6 million black turban periwinkles were collected. A cleaning operation was also carried out on 1,500 m² of intertidal area adjacent to the aquaculture facilities. In addition, a system for filtering the effluent from these facilities prevented the worm from being introduced into the open environment. It was considered eradicated in 1998, after five years of action. This success was made possible by several factors. The problem was recognised quickly, while the species was still limited in its spread. Furthermore, the parasite was highly specific to its host. In addition, the species' reproduction involved only a short pelagic larval phase, which restricted its dispersal. Finally, the biological cycle of the species was already known, and this made it possible to take the appropriate action.

The striped black mussel in Northern Australia

The second case of eradication involved the black-striped mussel (*Mytilopsis sallei*) in the marina in Darwin, Australia, in 1999 (Willan *et al.*, 2000). Originally from Central and South America, the species was known to have been invasive, particularly in Fiji and Singapore. Detected in March 1999 in high densities (around 24,000 individuals/m²) by divers in charge of the IAS monitoring network, the species had colonised several port infrastructures within one of the Cullen Bay marinas. Because of the tidal range in this sector (> 8 m), the marinas are equipped with locks that provide total isolation from the ocean environment. Within a week, the presence of this mussel was confirmed in the port's three marinas, as was its absence outside the port. The initial introduction is assumed to have taken place in the previous six months. Given the environmental risks of proliferation and impacts on the pearl farming industry and aquaculture infrastructures, a rapid response was implemented. A state of natural disaster was declared, and the marinas were quarantined. The 220 boats present were also quarantined. The eradication campaign was carried out using chemical water treatment with chlorine (187 tonnes of bleach/sodium hypochlorite) and copper sulphate (7.5 tonnes, 10 ppm) in the three marinas over a two-week period, at a cost of AUS$2.2 million. The aim is to destroy all forms of life, including exotic ones, before allowing natural renaturation. The bleach treatment proved ineffective, as it was diluted by seasonal rains, whereas the copper sulphate treatment remained effective. Four tonnes of dead fish were collected. The biofouling on the hulls of the boats and the water circuits of their engines were also treated, the latter by disinfection with boiling water, copper sulphate and detergents. The boats were dried out and careened in boiling water to prevent any further transfer, and 760 other boats that had previously been moored in the three marinas were inspected and treated. All the mussels and a large part of the marine life had been destroyed by 18 April of the same year. Following these treatments, the impact of which has not been assessed, marine life gradually repopulated the marinas. This eradication prevented economic losses estimated at $200 million and significant environmental damage to the coastal zone of Northern Australia. Since then, preventive inspection systems have been imposed on international boats to avoid any further introductions. This biovigilance avoids the need for drastic measures that are neither desirable nor advisable, such as the use of large volumes of chemical substances that could have significant impacts. In this case study, the success of eradication depended on the existence of a dedicated monitoring network and rapid action protocols. The most significant action was the total isolation of the marinas thanks to the locks. The most important outcome of this operation is twofold: it has raised public awareness and set in place a long-term process for strengthening surveillance of the Australian coastline for IAS.

Caulerpa in California, United States

Detected in June 2000 in the Agua Hedionda lagoon and Huntington Harbour, California, USA, the invasive alga *Caulerpa taxifolia* was already known for its impact during the fifteen years of its invasion of the Mediterranean Sea. Already listed in the US Federal Noxious Weed List in 1999, a consensus was quickly reached by all the managers to set up a task force, the Southern California Caulerpa Action Team (SICCAT). Treatments and field checks began 17 days after the first detection and continued until the species disappeared at the end of 2002 (Anderson, 2005).

Three well-integrated components of the rapid action plan help to explain the success of this eradication: expertise and prior knowledge of the biology of the algae, in-depth knowledge of the site and local customs, and the mobilisation of divers already operating on site. The rapid provision of financial resources (US$1.2 million per year) ensured the success of the operation, which consisted of regularly covering the algae with tarpaulin-covered PVC structures into which 12% bleach was injected in liquid form or in tablet form for the smallest areas. The areas treated varied from 1 to 500 m² for a total surface area of 2,200 m². Monitoring of the sites continued until 2005. Genetic analyses subsequently showed that the strain of algae was identical to that found in the Mediterranean Sea, with the initial introduction most likely to have come from a boat that had previously travelled in the Mediterranean and whose home port/cargo base was this very port in California. Since then, the possession, sale, and transport of *Caulerpa* have been prohibited under California's 2001 law, which was further reinforced by a total ban in 2023. Vigilance is still called for, as another alien *Caulerpa* species, *C. prolifera*, originating from Florida (USA), was detected in Southern California in 2021 and again in 2023, followed and controlled by the same treatment procedures.

These three examples, which could be considered as counter-examples given the current overall situation, have certain features in common that explain their success: knowledge of the characteristics of the species at both individual and population levels, effective consideration of scientific recommendations based on local and international knowledge, and financially supported management, detection and rapid action procedures, even before an invasion gets out of control. This will ensure that monitoring and prevention systems are put in place to maintain the good ecological status of the ecosystems concerned.

HOW CAN RESEARCH AND EXPERTISE CONTRIBUTE?

As well as simply producing knowledge, research in this field aims to support public policy and provide the most accurate information possible, so that decisions can be made and management organised quickly and effectively. It is essential that this information is scientifically based. For an issue that is global in scope, scientific expertise requires exchanges and databases on an international scale. The scientific community is organising itself accordingly: dedicated scientific journals supported by the Invasivesnet network — an international association for open knowledge on invasive alien species[26] — facilitate these exchanges, complemented by international scientific conferences such as the International Conference on Aquatic Invasive Species (ICAIS) and international working groups such as those of the ICES and the IUCN, with the treatment of concrete cases to identify research priorities, best practices, decision-making tools and management recommendations. This global scientific momentum is reinforced by the storage and organisation of data in international information systems and databases (e.g. AquaNIS, EASIN) (Magliozzi *et al.*, 2024).

As shown in **Figure 1**, research and development activities can be found at every stage of the biological invasion process.

ASSESSING RISKS

As prevention remains the most effective management option in the marine environment, many research projects aim to improve the detection of alien species and to identify the risks for potential places of introduction (e.g. ports) in order to anticipate them. Predicting which alien species will have a negative impact on local biodiversity has long been a research priority.

---

26. https://www.reabic.net/journals/

Several risk assessment and decision support methods have been developed, such as the AS-ISK application already mentioned, which incorporates climate change criteria into its assessment (Copp *et al.*, 2016; Vilizzi *et al.*, 2021). AS-ISK is based on 55 questions (ranging from biology to known impacts) that require answers derived from validated scientific data (e.g. publications), but which also call on expert knowledge ('expert opinion') through surveys, and on their level of confidence by taking uncertainties into account. Overall, risk assessment tools establish relative scores that can be used to rank priority species. For example, they are used to identify species that require listing under national and European legislation in accordance with the 2014 regulation (Pisanu *et al.*, 2020).

Risk assessment approaches have been specially designed to take account of maritime transport, which circumvents natural biogeographical barriers. Knowing the history of ships and their specific characteristics is therefore essential in the risk assessment process. Depending on whether the vessel is a ro-ro ferry, a bulk carrier or a container ship, the risks will be different both in terms of hull surface area colonised and residence time in port, where longer periods are more likely to facilitate colonisation by biofouling (Hegele-Drywa *et al.*, 2024). Similarly, transit times from port to port are another criterion to be taken into account for the survival of species present on hulls. It should also be noted that new practices, such as reducing speed in maritime transport, are the subject of cost/benefit analyses based on multifactorial approaches that consider biological, economic and social criteria for optimisation purposes. The key factor here is the modelling approach applied to all the criteria (Drolet *et al.*, 2017). Based on a 'world' cartography, analysis of the maritime traffic network and ship flows enables the ecoregions they cross to be identified. Additional information is collected automatically and is increasingly used by scientists to assess the pressures on the environment (data relating to ballast tanks and biofouling): the Automatic Identification System (AIS), for instance, is a system of automated exchanges of messages between ships by VHF radio that enables ships and traffic monitoring systems (in France, this is the regional operational surveillance and rescue centres, *Centres Régionaux Opérationnels de Surveillance et de Sauvetage*: CROSS) to find out the identity, status, position and route of ships in the navigation zone. It was initially set up for safety reasons, in particular to prevent collisions at sea. Tournadre (2014) also demonstrated that the use of satellite altimetry data can be used to draw up maps of maritime traffic density to complement the often-fragmentary data from the AIS. By cross-referencing this information with our knowledge of the invasive alien species present in these ecoregions and their respective environmental conditions, we can establish a probability-of-introduction score for a given port. Combined with the relative impacts of these species, this will then give an overall risk score. This approach can also be employed to identify hotspots where species are introduced.

Research is also focusing on technological aspects. The BWM Convention sets efficiency targets (criterion D2) for deballasting, defined by three parameters: fewer than 10 viable organisms per cubic metre of water ≥50 µm in size; fewer than 10 viable organisms per millilitre for organisms ≥10 µm; and concentrations of certain microbes relevant to public health (e.g. coliforms) below specified thresholds. These indicators are currently being reassessed, as they are based on the number of organisms (propagule pressure) and not on the resilience of the community of species present. New technologies are currently being tested in controlled environments (mesocosms) in order to assess the conditions for effectiveness and compliance with this D2 criterion. The results of the research are therefore tending to change the descriptors in order to guarantee the effectiveness of deballasting at sea and the treatment of these ballast tanks on board to reduce the risks of introduction. Scientists are also assessing the effects of the exception and exemption measures provided for in the BWM Convention (Outinen *et al.*, 2021). As a reminder, three cumulative criteria must be met for this exemption: the vessel must operate exclusively between specified ports, must only fill its ballast tanks with water from these specified ports, and its exemption request must be accompanied by a risk analysis. For example, this could apply to a ferry serving only one island, sailing between two specific, nearby ports. Modelling approaches are used to assess the methods of application and their effectiveness.

In ballast water management, there is no such thing as zero risk, but treatment aims to reduce it to the lowest possible level. Despite these constraints, Australia, whose monitoring and control methods rank among the best worldwide, has reported a 30% failure rate of vessels in ship inspections since 2019, with chlorine and ultraviolet treatment systems being the most commonly used. This demonstrates the limits to the effectiveness of ballast water and sediment treatment processes, particularly for organisms in encysted forms. The priorities are therefore to improve treatment techniques, particularly for ballast sediments, and to optimise monitoring strategies.

Although biofouling on boat hulls is currently addressed only through management recommendations, researchers are developing technologies to identify its nature, measure the affected surface area, and design targeted treatment tools. These approaches also aim to reduce the use of chemical substances that are harmful to the environment. Two- and three-dimensional optical and imaging approaches are being developed with automated analyses using artificial intelligence (Riley *et al.*, 2024). Complementary 'blue chemistry' projects are looking for active compounds with properties similar to natural antifouling substances or materials that can inhibit the formation of bacterial films (biofilms), the first stage in the colonisation of boat hulls.

IMPROVING THE DETECTION OF INVASIVE ALIEN SPECIES

Species management depends on precise identification. Traditionally based on morphological taxonomy, this process has been transformed by biomolecular ('-omics') techniques, which are providing significant new insights into exotic species. Today, these datasets are best supplemented with complementary and independent information on biogeography, ecology, cytogenetics, and reproduction and development. This integrative taxonomy approach offers a more comprehensive framework (Pante *et al.*, 2015; Nunes *et al.*, 2024).

Complementary methods based on environmental DNA (barcode, metabarcoding) make it possible to detect and inventory species in a given environment and to differentiate between those that are cryptic and/or in immature stages (e.g. larval stages). The RAPSODI scientific project, led by Ifremer, aims to develop and validate biomolecular tools for detecting the DNA of the veined rapa whelk, an exotic gastropod mollusc whose population is on the increase in the Pertuis Charentais, and the polyclad marine flatworm *Idiostylochus tortuosus* (also called *Postenterogonia orbicularis*), present in the Thau lagoon and Arcachon basin. To carry out targeted detection of these two species in the environment, real-time PCR-type tools were used, based on the sequencing of previously collected specimens. These two tools detect the presence of DNA from these two species in a seawater sample (Pépin, 2023). This genetic information is integrated into international databases such as GenBank, BOLD (Barcode of Life Data System) and/or Midori2, the contents of which require continuous improvements in order to be fully operational (Ratnasingham and Hebert, 2007; Leray *et al.*, 2022). Although relatively recent, these approaches have demonstrated their effectiveness and accuracy in describing taxonomic and genetic diversity (Couton *et al.*, 2021; 2022). These methods do, however, still face a number of challenges, particularly standardisation, but are making steady progress towards reducing their own limitations (Rishan *et al.*, 2023). They are increasingly incorporated into monitoring protocols, supporting new strategies thanks to their accuracy and cost-effectiveness.

ASSESSING THE RISKS AND IMPACTS FOLLOWING AN INTRODUCTION

Although prevention remains the priority approach, research projects are mainly conducted at stages when exotic species are already in their invasive phase. While these approaches are relatively late, they nonetheless provide useful knowledge in the event of an invasion by this species on other coastlines, or in other countries, for the benefit of the scientific community and managers.

As a preliminary step, it would seem essential to have appropriate metrics for assessing and quantifying the impacts resulting from these invasion processes, particularly to prioritise actions. Insufficient consideration is also given to how these impacts evolve over time, including the adaptation of ecosystems to such pressures (Strayer *et al.*, 2006). Various scientific methods have been developed for impact assessment. Several hypotheses explaining invasion processes focus on trophic relationships and interspecific competition. They consider that the most disruptive IAS use natural resources more efficiently than native species. 'Functional response' (the use of resources as a function of their availability) is a tool for assessing, quantifying and comparing the ecological impacts of IAS (impact = functional response × abundance) (Faria *et al.*, 2023). Other methods, such as stable isotopes, are used to study trophic interactions and define the respective trophic niches during the invasion process. Karlson *et al.* (2024) were able to demonstrate the existence of a vacant trophic niche in the Baltic Sea that explains the success of the Atlantic rangia clam (*Rangia cuneata*), initially from Mexico, at invading this area. In a similar way, Maric *et al.* (2016) studied the interactions of four exotic species colonising a marine protected area on the island of Lampedusa, in the Mediterranean: *Caulerpa cylindracea*, red sea plume (*Asparagopsis taxiformis*), the urchin crab and the sea hare (*Aplysia dactylomela*). The *Caulerpa* invasion increased the diversity of available prey, thus facilitating the development of other exotic species, with the crab feeding mainly on *Caulerpa* and the sea hare competing trophically with native herbivores. In this case study, there was major restructuring of ecosystem functioning, even though this was *a priori* protected, and then a 'self-maintenance' of the invasion process. Overall, it is the understanding of the dynamics of the invasion process and of the functioning of the host ecosystem that are the focus of major scientific projects, studied mainly through an ecosystem and modelling approach.

The spatial approach is also of interest. Research into biogeography and ecology is crucial for identifying and predicting invasion hotspots. In the case of exotic species already showing large multi-localised populations, molecular biology tools are essential to better understand and retroactively identify introduction routes and proliferation processes. The study of several molecular genetic markers from populations of the green alga *Ulva australis* in France, compared with reference data from international genetic databases including populations of this alga worldwide and from herbarium collections, has particularly demonstrated that its presence is the result of multiple introductions via mutually independent pathways. The study suggests that maritime traffic between Japan and France since the 19th century has been the source of several introductions: an introduction concomitant with that of the Pacific oyster in the 1970s cannot fully explain the current situation (Sauriau *et al.*, 2021). The analysis of genetic markers from systematic sampling of populations of a species on a global scale, supplemented by analyses of samples from historical collections, will thus provide a deeper understanding of the biogeographical origins of introduced exotic species, the vectors and routes of introduction — including associated human activities — and particularly the history of the processes involved.

The use of predictive modelling makes it possible to identify future distribution areas on the basis of climate trajectories, as well as so-called 'refuge' zones linked to specific climatic events, by coupling with hydrodynamic modelling. These refuges offer exotic species a means of delaying their expansion during periods of unfavourable environmental conditions. This approach makes it possible to establish priority areas for management by identifying high-risk zones with the aim of delaying their expansion (Krumhansl *et al.*, 2023). Dispersal and the identification of refuge areas have been successfully applied to the study of the colonisation and larval dispersal of invertebrates on the west and east coasts of the United States, as well as for colonial ascidians in Nova Scotia (Canada) (Di Bacco and Lowen, 2024). Similarly, the synergy between the various pressures on the environment must be taken into account by using modelling approaches that go beyond the process of just biological invasion. For example, the effects of climate change have facilitated the expansion of lionfish from the eastern to the western Mediterranean through ongoing tropicalisation and the altered connectivity of water bodies, as they have for the expansion of the Pacific oyster towards northern latitudes (Novi *et al.*, 2021; Anglès d'Auriac *et al.*, 2017).

When a biological invasion is detected at a late stage, it is often necessary to set up a multidisciplinary, multi-year research project. This was the case with the proliferation of the American blue crab on the French Mediterranean coast, which has had significant impacts. Understanding the ecology of the species in its new ecosystem and its impact on biodiversity and environmental services is therefore fundamental. In particular, the analysis of crab migrations between the coastal marine environment and lagoons or estuaries via the geolocation of individuals (tags for tracking or marking) is important for management purposes. Population stratification, details of migrations linked to gender, ontogeny and reproduction, trophic aspects and environmental tolerances (temperature, salinity) are all required information. For this case study, several biological traits have already been acquired (Hourdez et Boyer, 2023): sexual maturity is reached between 8 and 10 months in the Canet lagoon, and the larval phase lasts between 30 and 70 days, facilitating dispersal. It is known to suffer above 32°C and die at 40°C, and that its respiration rate increases from 12°C, but its optimum is 24°C. A female can produce 2 million eggs per clutch, and it can almost double in size, growing from 9 cm to more than 16 cm in a single moult. All these scientific results characterise the biological traits of the species in its new environment, data that is necessary for modelling and, ultimately, the implementation of targeted management actions.

Other research approaches directly focus on the genetic characteristics of invasive species (e.g. genetic introgression, hybridisation) for management purposes. Several scientific projects aim to use the biology of species to release sterile individuals into the environment or, in aquaculture, to use selected sterile or monosex lines. Chemically sterilised males of invasive and migratory sea lampreys (*Petromyzon marinus*) have been used effectively for a decade as a biocontrol method in the United States (Johnson *et al.*, 2024). Research using the molecular scissors technique (CRISPR-Cas9) is also being developed to obtain sterile fish (Smanski *et al.*, 2024). These approaches raise their own questions, for example about the impact of these genetic modifications on the physiology of individuals (associated effects), on the maintenance (or not) of the migratory behaviour of the fish concerned, on their reproductive behaviour, or even on the potential permanent and non-targeted impacts. Ultimately, all these factors raise ethical and scientific deontological questions about the development of such research projects, which require independent ethical approval assessments prior to any experimentation in an open environment.

The humanities and social sciences (HSS) make a major contribution to the issue of biological invasions, in particular by conducting studies on the direct and indirect costs associated with them (e.g. indicators in the InvaCost project), or on cost/benefit analyses to assess eradication measures (Diagne *et al.*, 2021; Zeni *et al.*, 2021). These approaches are particularly important in situations of complex governance, where objectives may prove to be conflicting. The previously mentioned case of the American blue crab in the Mediterranean Sea illustrates this, as does the fishery for the veined rapa whelk, an invasive alien species in the Black Sea (Demirel *et al.*, 2021). Management objectives must comply with the MSFD, which aims to achieve good ecological status according to EU guidelines, including the reduction of invasive alien species, while at the same time meeting social justice objectives for a fishery that economically benefits rural populations. In addition, the use of dredges as fishing gear is considered incompatible with the objectives of good environmental status. Research on the veined rapa whelk has highlighted the need for adaptive management and co-management, with all stakeholders, including bodies from neighbouring countries, being taken into account if the management options adopted are to be accepted.

Overall, these issues of governance arrangements, particularly for marine protected areas, are one of the key areas for research in the international context of marine biodiversity protection and the '30 by 30' objectives (30% of marine areas protected by 2030), with 10% under strong protection. The management of marine protected areas and assessment of their effectiveness by research are essential here, giving full consideration to the various pressures. In 2016, Giakoumi *et al.* found that only three conservation plans, out of 119 marine protected areas studied worldwide, took account of invasive alien species in their decisions and management plans. However, management decisions, such as which habitats are critical, would be modified if this was included. In addition to the economic aspects, socio-cultural approaches are also addressed, particularly in terms of public perceptions of IAS and environmental degradation (Kapitza *et al.*, 2019). These dedicated studies also make it possible to assess the willingness to pay for environmental rehabilitation projects. The transdisciplinary approaches that can be applied, including the human and social sciences and the combining of the various scientific themes, are in line with the scenario and modelling frameworks prescribed by IPBES, particularly the conceptual framework on the future of nature, Nature Futures Framework (NFF) (Pereira *et al.*, 2020; IPBES, 2023).

WHAT ARE THE EXPECTATIONS FOR RESEARCH IN THE SHORT AND MEDIUM TERM?

The most important objective for the study of marine biological invasions remains the ability to predict both the invasion process and the resulting impacts, whether these are impacts on the environment and/or on human activities. Improving this forecasting capacity is the main objective for optimising the quality of the management and control of biological invasions. This requires acquiring and sharing knowledge about the species involved across the international scientific community, particularly by strengthening information exchange networks and dedicated databases, while also drawing on the results of targeted scientific projects. In 2020, for example, such cooperation made it possible at the European level to draw up prospective scenarios for future biological invasions via a working group bringing together scientists, managers and decision-makers, non-governmental organisations and specialists in the issues raised by global change. In particular, the need to take into account the interactions between different scales, from regional to local (pressures and public policies), was highlighted in the development of scenarios relevant at a fine scale (Pérez-Granados *et al.*, 2023).

Research is necessary in this field to build up an adequate knowledge base to support public policy, but also to transfer information to wider society. Resource centres for IAS play a certain role in this. However, this is a relatively recent field of activity, based on just a few well-documented case studies over a long period of time. Observations relating to biological invasions are often made only at a late stage and on an *ad hoc* basis, without a spatialised sampling strategy, or over a long period of time, thus limiting the capacity for dynamic analysis of the process. In response, we can only stress the importance of setting up research observatories in this field and the need for a reinforced national monitoring strategy.

The responses of ecosystems to invasions are highly variable, not only in terms of resilience but also in space and time. Effects on the functioning of ecosystems and on the evolution of species can take decades, requiring observations over long periods of time, which is ill-suited to the current functioning of research. The uncertainties associated with scientific results are also a weakness when it comes to making rapid decisions on management methods. Risk assessments are often based on the integration of scientific data that has already been published and validated, complemented by expert opinion to address the frequent lack of information in this area and the uncertainty that characterises new invasion events. Research must develop dedicated projects to improve support for the development and implementation of public policies.

Nevertheless, it should be noted that the issues raised in the ecology of biological invasions are quite similar to those known for marine biodiversity (Goulletquer *et al.*, 2013), as follows:

- Make an inventory of all aspects of biodiversity and develop the tools and resources needed to describe it;
- Understand the evolutionary and ecological processes responsible for the variety, abundance and distribution of genes, populations, communities and ecosystems in space and time;

- Assess how biodiversity patterns influence the functioning of ecosystems and the provision of services produced by the environment, including relations with the non-living world, and the associated socio-economic benefits;
- Understand the factors of change and adaptive responses;
- Support the development of management systems to achieve biodiversity conservation objectives, including decision-support tools.

Research on biological invasions must also be placed within a broader context, particularly both as a cause and as a consequence of anthropogenic global change. Collaboration with climatologists and oceanographers is essential to support modelling efforts and the development of scenarios (US EPA, 2008; Canning-Clode, 2015). The evolution of ecosystems linked to this global change is a dimension that needs to be taken fully into account, particularly in terms of the expansions, range changes and population dynamics of exotic species currently present, phenomena that are expected to increase in the coming decades (Chan *et al.*, 2019; Hellmann *et al.*, 2008; Mainka and Howard, 2010).

For example, experiments have shown that high temperatures associated with long heat waves will affect the structure of communities of both native and exotic species. Short heatwaves will have more marked effects on communities dominated by exotic species, but long heatwaves preferentially weaken native species (Castro *et al.*, 2021). One of the four priorities proposed by Ricciardi *et al.* (2020) is to identify climate mitigation and adaptation strategies by understanding the potential synergistic effects of multiple concurrent stress factors, particularly climate change, on the establishment and impact of invasive alien species. Other priorities include the need for a more complete framework for predicting variation in the behaviour, abundance and interspecific interactions of exotic species as a function of receiving environments, as well as their impacts (Ricciardi *et al.*, 2020). The ability to detect and assess risks requires taxonomic skills that cannot be fully replaced by molecular techniques. However, there is a scarcity of higher education courses in this field, which is known as the 'taxonomic handicap' (Faugère and

Mauz-Harpin, 2013). The fourth recommendation is to step up international cooperation in the field of biosecurity, focusing on sites with a high risk of dispersal of invasive alien species, such as commercial ports.

At global and regional scales, there is a clear shortage of studies conducted jointly by specialists in biological invasions and experts working on marine protected areas. Approaches to conservation and impacts on ecosystem services and their maintenance, while improving the coupling of socio-economic models with those on biological invasions and their management, also require further development (Wonham and Lewis, 2009). The IPBES NFF takes into account the different perspectives and values associated with nature and can therefore be considered as a basis for future projects on biological invasions.

CONCLUSION
Invasive marine species: what does the future hold?

The United Nations' Agenda 2030 has set 17 very specific sustainable development goals, including the conservation of marine biodiversity. The management and control of marine biological invasions contribute directly to achieving this goal, but those relating to food security, sustainable economic development, climate change and human health and well-being are just as important. Integrated governance requires recognition of the interactions between these different areas to develop coordinated public policies with shared benefits. For example, the issue of marine biological invasions must be seen in the context of a regenerative blue economy, as described by the IUCN, which goes beyond the current framework of 'sustainability' (Le Gouvello and Simard, 2024). From a scientific perspective, the issue must also be approached heuristically, through a co-constructive process, in order to arrive at management options that are considered 'acceptable' within a limited timeframe, taking into account the often incomplete knowledge available and the conflicting interests at stake (conservation versus development) (Meinard *et al.*, 2022).

The Global Biodiversity Framework, established at COP 15 in Kunming, Montreal (2022), sets a target of a 50% reduction in new introductions of alien species by 2030. This very ambitious objective will require a multiplication of actions and a change of trajectory if France is to meet the target. This will necessarily involve strengthened approaches to prevention, predictive analyses of the risks of introduction according to geographical sectors and maritime transport characteristics, and a fully operational rapid detection strategy, both in mainland France and French overseas territories.

It should be emphasised that efforts to communicate scientific findings to the public are increasing significantly, while research institutions are placing greater emphasis on social and societal concerns. This recent development has a direct impact on the management of biological invasions.

Overall, taking invasive alien species into account is essential and needs to be reinforced at every level to achieve effective management, particularly within territorial strategies, maritime spatial planning instruments and authorities, and management plans for marine protected areas, including highly protected zones. In France, unlike in countries such as New Zealand and Australia, actions to monitor these species, prevent their introduction and survey their population dynamics are not given sufficient priority in marine protected area management plans, undermining the conservation of important habitats (Giakoumi *et al.*, 2016). However, a shift is now under way, reflected in the growing inclusion of such actions in marine protected area management plans, the design of biosecurity plans and awareness-raising programmes, and increasing numbers of managers and politicians being trained about the issue of biological invasions.

Despite these factors and the lack of scientific data, which is sometimes insufficiently consolidated or inaccessible, management actions and resources are available at the various stages of a biological invasion and can therefore be implemented. The IPBES (2023) stresses that ambitious progress in the management of biological invasions can be achieved by adopting an integrated governance approach that defines strategic actions. These actions can be defined in seven points:

- increased collaboration and coordination of international and regional mechanisms;
- the development of appropriate national strategies;
- shared commitment and understanding of the specific roles of the players;
- improving the coherence of public policies;
- facilitating the involvement of the different players, from government bodies to the general public;

- supporting and mobilising resources for innovation, research and appropriate technologies;
- support for information systems, infrastructures and information sharing.

At European and national levels, significant progress has been made in recent years. In particular, the 2014 EU regulation adequately defines the various actions that warrant full consideration. However, there is still room for improvement in the way this regulatory text is implemented.

The lionfish invasion in the Mediterranean Sea was the subject of dedicated monitoring, a regional action plan and a risk assessment for the period 2016–2021. Although its negative effects (current and future) on marine biodiversity are known, the process of including it on the European list of species of concern had still not been finalised in September 2025. This highlights the need to better address the issues at stake and to shorten the time between the production of consolidated scientific assessments and their consideration by managers. These assessments may include uncertainties due to limited data, such as incomplete monitoring, but they must still inform the implementation of rapid action plans. The procedures will take a long time to be consistent with the objectives set, which *de facto* prioritises long-term management. A number of limitations have been identified that are also due to the differences between terrestrial and marine issues, and in the case of the latter, the connectivity between ecosystems is much more pronounced, making timely decision-making essential (Kleitou *et al.*, 2021).

An additional complexity in drawing up lists at the European level is that some native species may have 'vulnerable' status in some countries, but invasive status in others (Baquero *et al.*, 2023). At this stage, these species are 'outside the scope' of European regulations in terms of drawing up the list of species of concern and represent a real challenge in terms of managing and preserving their genetic diversity. This paradox, which pits the need for conservation in the native range against the need for reduction where the invasion is developing, concerns 317 taxa, 17 of which are marine (16 invertebrates and 1 brown alga) and mostly located in the Mediterranean. More than a quarter of these have populations with 'threatened' status in their area of origin. This also raises the issue of drawing up single national lists, as is the case in France, despite the existence of geographical areas with very different levels of biodiversity. Increasingly, Mediterranean species (or species introduced into the Mediterranean) are surviving in the southern Bay of Biscay, reflecting ongoing environmental changes. The way to manage these difficult cases could potentially involve clarification of national lists and greater cooperation between scientists, managers and politicians when the situation is international. Article 11 of the 2014 European regulations offers such a possibility:

> At the request of the Member States involved, the Commission shall facilitate cooperation and coordination. If certain conditions are met — including impacts on biodiversity, ecosystem services, human health or the economy and a thorough justification by the requesting states — the Commission may, by implementing acts, require those Member States to apply, *mutatis mutandis*, specific Articles of the Regulation in their territory or part of it.

Note that such enhanced cooperation can be beneficial, particularly as an opportunity to exchange information generated in impacted areas to develop new conservation approaches in areas of origin (Gibson and Yong, 2017).

The implementation of this regulation is underway at a national level. France has adopted action plans in this area, particularly by drawing up lists of species for France and its overseas territories and departments, as well as an initial risk analysis for a certain number of marine species and the ongoing updating of the Environment Code. For the marine environment, the monitoring system needs to be strengthened, as does the implementation of rapid eradication plans. However, cost–benefit analyses are still needed, particularly to consider the costs of maintaining or restoring the functionality of the ecosystems affected. The MSFD, with its objective of achieving good ecological status, is the priority instrument for making these advances and for developing national coordination in France, whose actions would need to be extended to the overseas territories.

Given the specific environmental characteristics of the marine environment, the main difficulty in implementing the European regulation concerns the restoration of altered ecosystems (art. 20, 2014 regulation). The adoption in July 2024 of the European Nature Restoration Regulation (EU) 2024/1991 is an important step in strengthening the dedicated 2014 regulation by setting quantified targets for the ecosystems concerned. The aim of this regulation is to restore nature and ecosystems to a state of good conservation. It obliges EU countries to draw up national restoration plans and sets a binding European target for effective restoration measures to cover at least 20% of the land and sea area of the EU by 2030. By 2050, measures must be put in place for all ecosystems in need of restoration. Implementing this regulation will require precise planning to identify and prioritise the marine ecosystems in need of restoration, while considering the positions of stakeholders and the conflicts of use that are bound to arise.

At the international level, the IMO is tightening its regulations on the main vector of introduction, namely maritime transport. It should be emphasised once again that the BWM Convention undeniably represents one of the major advances in controlling the introduction of exotic species, with clear benefits for both environmental and public health. It was and is a source of technological innovation for the development of *ad hoc* ballast water treatment technologies. Scientific work is aimed at improving the effectiveness of ballast water treatment by focusing on optimised treatment systems, criteria (e.g. sediments) and the attainment of standardised objectives. In addition to the implementation of this agreement,[27] priority is given to the prevention and treatment of marine biofouling. The scientific projects in this area will contribute in the short and medium term to the development of innovative, environmentally friendly technologies to improve the control and monitoring of the introduction of exotic species, thereby reducing the problem of marine biological invasions.

---

27. https://www.imo.org/en/OurWork/PartnershipsProjects/Pages/GloFouling-Project.aspx

References

Anderson L.W.J., 2005. California's reaction to *Caulerpa taxifolia*: a model for invasive species rapid response. *Biological Invasions*, 7, 1003–1016. https://doi.org/10.1007/s10530-004-3123-z

Anglès d'Auriac M.C., Marc B., Rinde E., Norling P., Lapègue S. *et al.*, 2017. Rapid expansion of the invasive oyster *Crassostrea gigas* at its northern distribution limit in Europe: naturally dispersed or introduced? *PLoS ONE*, 12 (5), e0177481. https://doi.org/10.1371/journal.pone.0177481

Arias A., Richter A., Anadon N., Glasby C.J., 2013. Revealing polychaetes invasion patterns: identification, reproduction and potential risks of the Korean ragworm, *Perinereis linea* (Treadwell), in the Western Mediterranean. *Estuarine, Coastal and Shelf Science*, 131, 117–128. https://doi.org/10.1016/j.ecss.2013.08.017

Bacher S., Blackburn T.M., Essl F., Genovesi P., Heikkilä J. *et al.*, 2018. Socio-economic impact classification of alien taxa (SEICAT). *Methods in Ecology and Evolution*, 9 (1), 159–168. https://doi.org/10.1111/2041-210X.12844

Baquero R.A., Oficialdegui F.J., Ayllón D., Nicola G.G., 2023. The challenge of managing threatened invasive species at a continental scale. *Conservation Biology*, 37, c14165, 9 p. https://doi.org/10.1111/cobi.14165

Barillé L., Le Bris A., Méléder V., Launeau P., Robin M. *et al.*, 2017. Photosynthetic epibionts and endobionts of Pacific oyster shells from oyster reefs in rocky versus mudflat shores. *PLoS ONE*, 12 (9), e0185187. https://doi.org/10.1371/journal.pone.0185187

Bellard C., Bernery C., Leclerc C., 2021. Looming extinctions due to invasive species: irreversible loss of ecological strategy and evolutionary history. *Global Change and Ecology*, 27 (20), 4947–5403. https://doi.org/10.1111/gcb.15771

Biodiversa, 2017. Policy Brief. Action on invasive alien species should better anticipate climate change effects on biological invasions in Europe, 4 p. https://www.biodiversa.eu/wp-content/uploads/2022/12/policy_brief_invasive-alien-species.pdf

Boissin E., Neglia V., Baksay S., Micu D., Bat L. *et al.*, 2020. Chaotic genetic structure and past demographic expansion of the invasive gastropod *Tritia neritea* in its native range, the Mediterranean Sea. *Nature, Scientific Reports*, 10, 21624. https://doi.org/10.1038/s41598-020-77742-3

Bottacini D., Pollux B.A.J., Nijland R., Jansen P.A., Naguib M., Kotrschal A., 2024. Lionfish (*Pterois miles*) in the Mediterranean Sea: a review of the available knowledge with an update on the invasion front. *NeoBiota*, 92, 233–257. https://doi.org/10.3897/neobiota.92.110442

Boudouresque C.F., Bernard G., Bonhomme P., Charbonnel E., Diviacco G. *et al.*, 2012. Protection and conservation of *Posidonia oceanica* meadows. RAMOGE and RAC/SPA, 202 p. https://hal.science/hal-00808491

Booy O. *et al.*, 2020. Using structured eradication feasibility assessment to prioritize the management of new and emerging invasive alien species in Europe. *Global Change Biology*, 26 (11), 6235–6250. https://doi.org/10.1111/gcb.15280

Cadée G.C., 2001. Herring gulls learn to feed on a recent invader the Dutch Wadden Sea, the Pacific oyster *Crassostrea gigas*. *Basteria*, 65 (1/3), 33–42. https://natuurtijdschriften.nl/pub/597203

Caldow R.W., Stillman R.A., dit Durell S.E., West A.D., McGrorty S. *et al.*, 2007. Benefits to shorebirds from invasion of a non-native shellfish. *Proc R Soc Biol Sci Ser B.*, 274, 1449–1455.

Canning-Clode J. (ed.), 2015. *Biological Invasions in Changing Ecosystems. Vectors, Ecological Impacts, Management and Predictions*, De Gruyter Open Ltd, 473 p.

Carlton J.T., 1979. History, biogeography and ecology of the introduced marine and estuarine invertebrates of the Pacific coast of North America. PhD. Dissertation, University of California, 904 p.

Carlton J.T., 1996. Biological invasions and cryptogenic species. *Ecology*, 77 (6), 1653–1655.

Carlton J.T., 1999. Molluscan invasions in marine and estuarine communities. *Malacologia*, 41 (2), 439–454.

Carlton J.T., Hodder J., 1995. Biogeography and dispersal of coastal marine organisms: experimental studies on a replica of a 16th century sailing vessel. *Marine Biology*, 121, 721–730. https://doi.org/10.1007/BF00349308

Castellanos-Galindo G.A., Robertson D.R., Torchin M.E., 2020. A new wave of marine fish invasions through the Panama and Suez Canals. *Nature Ecology and Evolution*, 4, 1444–1446.

Castro N., Ramalhosa P., Cacabelos E., Costa J.L., Canning-Clode J. *et al.*, 2021. Winners and losers: prevalence of non-indigenous species under simulated marine heatwaves and high propagule pressure. *Marine Ecology Progress Series*, 668, 21–38.

CBD, 2022. Cadre Mondial de la biodiversité de Kunming-Montéral post 2020. CBD/COP/DEC/15/4, 16 p. https://www.cbd.int/decisions/cop?m=cop-15

Chan F.T., Stanislawczyk K., Sneekes A.C., Dvoretsky A., Gollasch S. *et al.*, 2019. Climate change opens new frontiers for marine species in the Arctic: current trends and future invasion risks. *Global Change Biology*, 25, 25–38. https://doi.org/10.1111/gcb.14469

Chanet B., Desoutter-Meniger M., Bogorodsky S.V., 2012. Range extension of Egyptian sole *Solea aegyptiaca* (Soleidae: Pleuronectiformes), in the Red Sea. *Cybium*, 36 (4), 581–584.

Comité français de l'UICN, 2018. La valorisation socio-économique des espèces exotiques envahissantes établies en milieux naturels : un moyen de régulation adapté. Première analyse et identification des points de vigilance, 82 p.

Comité français de l'UICN, 2019. Espèces exotiques envahissantes marines : risques et défis pour les écosystèmes marins et littoraux des collectivités françaises d'outre-mer. État des lieux et recommandations. Paris, France, 100 p.

Copp G.H., Vilizzi L., Tilburyd H., Stebbingd P., Tarkanc A.S. *et al.*, 2016. The Aquatic Species Invasiveness Screening Kit (AS-ISK): a generic risk identification tool for marine, brackish and freshwater taxa. *Management of Biological Invasions*, 7 (4), 343–350.

Courchamp F., Fournier A., Bellard C., Bertelsmeier C., Bonnaud E. *et al.*, 2017. Invasion biology: specific problems and possible solutions. *Trends in Ecology and Evolution*, 32 (1), 13–22. https://doi.org/10.1016/j.tree.2016.11.001

Couton M., Baud A., Daguin-Thiébaut C., Corre E., Comtet T. *et al.*, 2021. High-throughput sequencing on preservative ethanol is effective at jointly examining infraspecific and taxonomic diversity, although bioinformatics pipelines do not perform equally. *Ecology and Evolution*, 11, 5533–5546. https://doi.org/10.1002/ece3.7453

Couton M., Lévêque L., Daguin-Thiébaut C., Comtet T., Viard F., 2022. Water eDNA metabarcoding is effective in detecting non-native species in marinas, but detection errors still hinder its use for passive monitoring. *Biofouling*, 38 (4), 367–383. https://doi.org/10.1080/08927014.2022.2075739

Craig M.T., Burke J., Clifford K., Mochon-Collura E., Chapman J.W. *et al.*, 2018. Trans-Pacific rafting in tsunami associated debris by the Japanese yellowtail jack *Seriola aureovittata* (Temminck & Schlegel, 1845). *Aquatic Invasions*, 13 (1), 173–177.

Culver C.S., Kuris A.M., 2000. The apparent eradication of a locally established introduced marine pest. *Biological Invasions*, 2, 245–253.

Demirel N., Ulman A., Yildiz T., Ertör-Akyazi P., 2021. A moving target: achieving good environmental status and social justice in the case of an alien species, Rapa whelk in the Black Sea. *Marine Policy*, 132, 104687. https://doi.org/10.1016/j.marpol.2021.104687

DeRoy E., Scott R., Hussey N., MacIsaac H., 2020. High predatory efficiency and abundance drive expected ecological impacts of a marine invasive fish. *Marine Ecology Progress Series*, 637, 195–208. https://doi.org/10.3354/meps13251

Diagne C., Leroy B., Vaissière A.-C., Gozlan R.E., Roiz D. *et al.*, 2021. High and rising economic costs of biological invasions worldwide. *Nature*, 592, 571–576. https://doi.org/10.1038/s41586-021-03405-6

DiBacco C., Lowen J.B., 2024. Range expansion of coastal marine non-indigenous species into thermal refuge habitat can be mediated by environmental variability in changing coastal environments. Abstract 71, ICAIS 2024.

Diaz *et al.*, 2015. The IPBES Conceptual Framework: connecting nature and people. *Current Opinion in Environmental Sustainability*, 14, 1–16. http://dx.doi.org/10.1016/j.cosust.2014.11.002

Drolet D., DiBacco C., Locke A., McKenzie C.H., McKindsey C.W. *et al.*, 2017. Optimizing screening protocols for non-indigenous species: are currently used tools over-parameterized? *Management of Biological Invasions*, 8 (2), 171–179. https://doi.org/10.3391/mbi.2017.8.2.05

Edgell T., Hollander J., 2011. The evolutionary ecology of European green crab (*Carcinus maenas*) in North America. *In: In the Wrong Place: Alien Marine Crustaceans. Distribution, Biology and Impacts* (B.S. Galil, P.F. Clark, J.T. Carlton, eds), Dordrecht, Springer, 641–659.

El-Serehy H.A., Abdallah H.S., Al-Misned F.A., Irshrad R., Al-Farraj S.A. *et al.*., 2018. Aquatic ecosystem health and trophic status classification of the Bitter Lakes along the main connecting link between the Red Sea and the Mediterranean. *Saudi Journal of Biological Sciences*, 25, 204–221.

Epstein G., Smale D.A., 2017. *Undaria pinnatifida:* a case study to highlight challenges in marine invasion ecology and management. *Ecology and Evolution*, 7, 8624–8642.

Essink K., Oost A.P., 2019. How did *Mya arenaria* (Mollusca: bivalvia) repopulate European waters in mediaeval times? *Marine Biodiversity*, 49, 1–10.

Essl F., Dullinger S., Genovesi P., Hulme P.E., Jeschke J.M. *et al.*., 2019. A conceptual framework for range-expanding species that track human-induced environmental change. *Bioscience*, 69 (11), 908–919.

FAO, 2024. *Fishery and Aquaculture Statistics: Yearbook 2021*, FAO Yearbook of Fishery and Aquaculture Statistics, Rome. https://doi.org/10.4060/cc9523en

Faria L., Cuthbert R.N., Dickey J.W.E., Jeschke J.M., Ricciardi A. *et al.*, 2023. The rise of the functional response in invasion science: a systematic review. *NeoBiota*, 85, 43–79. https://doi.org/10.3897/neobiota.85.98902

Faugère E., Mauz-Arpin I., 2013. Une introduction au renouveau de la taxonomie. *Revue d'anthropologie des connaissances*, V.7/2, 349–364. http://journals.openedition.org/rac/6506

Font T., Gill J., Lloret J., 2018. The commercialization and use of exotic baits in recreational fisheries in the north-western Mediterranean: environmental and management implications. *Aquatic Conservation: Marine and Freshwater Ecosystems*, 28, 651–661. https://onlinelibrary.wiley.com/doi/10.1002/aqc.2873

Fraser-Clark K., 2024. Invasive green crab (*Carcinus maenas*) are vectors for pathogens of the American lobster (*Homarus americanus*). *In: ICAIS Conference*, May 2024.

Galil B., Froglia C., Noël P., 2002. *CIESM Atlas of Exotic Species in the Mediterranean. V.2.2: Crustaceans: Decapods and Stomatopods*, CIESM Publishers (International Commission for the Scientific Exploration of the Mediterranean Sea), Monaco, 192 p. http://www.ciesm.org/atlas/appendix2.html

Galil B. *et al.*, 2015. The enlargement of the Suez Canal and introduction of non-indigenous species to the Mediterranean Sea. *ASLO*, 2–4.

Galil B.S., Marchini A., Occhipinti-Ambrogi A., 2018. East is east and West is west? Management of marine bioinvasions in the Mediterranean Sea. *Estuarine, Coastal and shelf Science*, 201, 7–16.

Galil B., Danovaro R., Rothman S.B., Gevili R., Goren M., 2019. Invasive biota in deep-sea Mediterranean: an emerging issue in marine conservation and management. *Biological Invasions*, 21, 2, 281–288. https://doi.org/10.1007/s10530-018-1826-9

García-Gómez J.C., Florido M., Olaya-Ponzone L., Sempere-Valverde J., Megina C., 2021a. The invasive macroalga *Rugulopteryx okamurae*: substrata plasticity and spatial colonization pressure on resident macroalgae. *Frontiers in Ecology and Evolution*, 9, 631754. https://doi.org/10.3389/fevo.2021.631754

García-Gómez J.C., Garrigos M., Garrigos J., 2021b. Plastic as a vector of dispersion for marine species with invasive potential. A review. *Frontiers in Ecology and Evolution*, 9, 629756. https://doi.org/10.3389/fevo.2021.629756

Giakoumi S., Guilhaumon F., Kark S., Terlizi A., Claudet J. *et al.*, 2016. Space invaders: biological invasions in marine conservation planning. *Diversity and Distributions*, 22, 1220–1231.

Gibson I., Yong D.L., 2017. Saving two birds with one stone: solving the quandary of introduced, threatened species. *Frontiers in Ecology and the Environment*, 15 (1), 35–41. https://doi.org/10.1002/fee.1449

GISD, 2024. Global Invasive Species Database. https://www.iucngisd.org/gisd/

Gomez-Chiarri M., Cobb J.S., 2012. Shell disease in the American Lobster, *Homarus americanus*: a synthesis of research from the New England Lobster research initiative: lobster shell disease. *Journal of Shellfish Research*, 31 (2), 583–590. https://doi.org/10.2983/035.031.0219

Goulletquer P., 2016. *Guide des organismes exotiques marins*, Éditions Belin, 305 p.

Goulletquer P., 2022. Espèces exotiques et transport maritime. *In: IVe Congrès du GIS d'histoire et sciences de la mer*, Nice, 18–20 May 2022.

Goulletquer P., Héral M., 1997. Marine molluscan production trends in France: from fisheries to aquaculture. *NOAA Tech. Rep. NMFS*, 129, 137–164.

Goulletquer P., Lacroix D., 2022. Aquaculture et biodiversité à 2050. *Futuribles*, 447, 65–77.

Goulletquer P., Gros P., Bœuf G., Weber J., 2013. *Biodiversité en environnement marin*, Éditions Quæ, 207 p.

Grizel H., Héral M., 1991. Introduction into France of the Japanese oyster (*Crassostrea gigas*). *Journal du Conseil/Conseil permanent international pour l'exploration de la mer*, 47, 399–403.

Gruet Y., Héral M., Robert J.M., 1976. Premières observations sur l'introduction de la faune associée au naissain d'huîtres japonaises *Crassostrea gigas* (Thunberg), importé sur la côte atlantique française. *Cahiers de biologie marine*, 17, 173–184.

Hacker S.D., Dethier M.N., 2009. Differing consequences of removing ecosystem-modifying invaders: significance of impact and community context to restoration potential. *In: Biological Invasions of Marine Ecosystems* (G. Rilov, J.A. Crooks, eds). *Ecological Studies*, 204, 375–385.

Hegele-Drywa J., Normant-Saremmba M., Wojcik-Fudalewska D., 2024. Small sea with high traffic: what is the biofouling potential of commercial ships in the Baltic Sea. *Biofouling*, 40 (3–4), 280–289. https://doi.org/10.1080/08927014.2024.2353025

Hellmann J.J., Byers J.E., Bierwagen B.G., Dukes J.S., 2008. Five potential consequences of climate change for invasive species. *Conservation Biology*, 22 (3), 534–543. https://doi.org/10.1111/j.1523-1739.2008.00951.x

Henry M., Leung B., Cuthbert R.N. *et al.*, 2023. Unveiling the hidden economic toll of biological invasions in the European Union. *Environmental Sciences Europe*, 35, 43. https://doi.org/10.1186/s12302-023-00750-3

Herbert R.J.H., Davies C.J., Bowgen K.M., Hatton J., Stillman R.A., 2018. The importance of nonnative Pacific oyster reefs as supplementary feeding areas for coastal birds on estuary mudflats. *Aquatic Conservation: Marine and Freshwater Ecosystems*, 28 (6), 1294–1307. https://doi.org/10.1002/aqc.2938

Hourdez S., Boyer T., 2023. Biologie de l'espèce *C. sapidus* dans l'étang du Canet (parasitologie, suivi de la reproduction). Conférence interrégionale « Actions de connaissance et de gestion du Crabe Bleu ». Pôle-relais lagunes méditerranéennes. Plan d'action régional Crabe. https://pole-lagunes.org

Huvet A., Lapègue S., Magoulas A., Boudry P., 2000. Mitochondrial and nuclear DNA phylogeography of *Crassostrea angulata*, the Portuguese oyster endangered in Europe. *Conservation Genetics*, 1 (3), 251–262. https://doi.org/10.1023/A:1011505805923

Ingeman K.E., 2016. Lionfish cause increased mortality rates and drive local extirpation of native prey. *Marine Ecology Progress Series*, 558, 235–245. https://doi.org/10.3354/meps11821.

ICES, 2004. Alien species alert: *Rapana venosa* (veined whelk) (R. Mann *et al.*, eds). *ICES Cooperative Research Report*, (264), 14 p.

IPBES, 2019. Global assessment report on biodiversity and ecosystem services of the Intergovernmental Science-Policy Platform on Biodiversity and Ecosystem Services (E.S. Brondizio *et al.*, eds). IPBES secretariat, Bonn, Germany, 1 148 p. https://doi.org/10.5281/zenodo.3831673

IPBES, 2023. Summary for policymakers of the thematic assessment report on invasive alien species and their control of the Intergovernmental Science-Policy Platform on Biodiversity and Ecosystem Services (H.E. Roy *et al.*, eds). IPBES secretariat, Bonn, Germany. https://doi.org/10.5281/zenodo.7430692

IUCN, 2020. IUCN EICAT categories and criteria. The environmental impact classification for Alien taxa. International Union for Conservation of Nature. https://iucn.org/resources/publication/iucn-eicat-categories-and-criteria-first-edition

Jaric I., Heger T., Monzon F.C., Jeschke J.M., Kowarik I. *et al.*, 2019. Crypticity in biological invasions. *Trends in Ecology and Evolution*, 34 (4), 291–302. https://doi.org/10.1016/j.tree.2018.12.008

Johnson N.S., Lewandoski S.A., Jubar A.K., Symbal M.J., Solomon B.M. *et al.*, 2024. A decade-long study demonstrates that a population of invasive sea lamprey

(*Petromyzon marinus*) can be controlled by introducing sterilized males. *Nature Scientific Reports*, 14, 12689. https://doi.org/10.1038/s41598-024-61460-1

Jongma D.N., Campo D., Dattolo E., D'Esposito D., Duchi A. *et al.*, 2013. Identity and origin of a slender *Caulerpa taxifolia* strain introduced into the Mediterranean Sea. *Botanica Marina*, 56 (1), 27–39. https://www.degruyterbrill.com/document/doi/10.1515/bot-2012-0175/html

JRC, 2024. Joint Research Center. https://commission.europa.eu/about-european-commission/departments-and-executive-agencies/joint-research-centre_en

Kapitza K., Zimmermann H., Martin-Lopez B., von Wehrden H., 2019. Research on social perception of invasive species: a systematic literature review. *Neobiota*, 43, 47–68. https://www.sciencedirect.com/org/science/article/pii/S1619003319000483

Karlson A.M., Kautsky N., Granberg M., Garbaras A., Lim H. *et al.*, 2024. Resource partitioning of a Mexican clam in species-poor Baltic Sea sediments indicates the existence of a vacant trophic niche. *Nature Scientific Reports*, 14, 12527. https://doi.org/10.1038/s41598-024-62832-3

Katsanevakis S., Wallentinus I., Zenetos A., Leppäkoski E., Cimar M.E. *et al.*, 2014. Impacts of invasive alien marine species on ecosystem services and biodiversity: a pan-European review. *Aquatic Invasions*, 9, 391–423. https://www.reabic.net/aquaticinvasions/2014/issue4.aspx

Kiessling T., Gutow L., Thiel M., 2015. Marine litter as habitat and dispersal vector. *In: Marine Anthropogenic Litter* (M. Bergman, L. Gutow, M. Klages, eds), Springer International Publishing, Cham, 141–181. https://doi.org/10.1007/978-3-319-16510-3_6

Kindinger T.L., Albins M.A., 2017. Consumption and non-consumption effects of an invasive marine predator on native coral-reef herbivores. *Biological Invasions*, 19 (1), 131–146.

Klein J., Verlaque M., 2001. The *Caulerpa racemosa* invasion: a critical review. *Marine Pollution Bulletin*, 56, 2, 205–225. https://www.sciencedirect.com/science/article/pii/S0025326X07003591

Kleitou P., Hall-Spencer J.M., Savva I. *et al.*, 2021. The case of Lionfish (*Pterois miles*) in the Mediterranean Sea demonstrates limitations in EU legislation to address marine biological invasions. *Journal of Marine Science and Engineering*, 9 (3), 325. https://doi.org/10.3390/jmse9030325

Kochmann J., Buschbaum C., Volkerbon N., Reise K., 2008. Shift from native mussels to alien oysters: differential effects of ecosystem engineers. *Journal of Experimental Marine Biology and Ecology*, 364 (1), 1–10. https://doi.org/10.1016/j.jembe.2008.05.015

Kourantidou M., Haubrock P.J., Cuthbert R.N., Bodey T.W., Lenzner B. *et al.*, 2022. Invasive alien species as simultaneous benefits and burdens: trends, stakeholder perceptions and management. *Biological Invasions*, 24, 1905–1926 https://doi.org/10.1007/s10530-021-02727-w

Krumhansl K., Gentleman W., Lee K., Ramey-Balci P., Goodwin J. *et al.*, 2023. Permeability of coastal biogeographic barriers to marine larval dispersal on the east and west coasts of North America. *Global Ecology and Biogeography*, 32 (6), 945–96 https://doi.org/10.1111/geb.13654

Labrune C., Amilhat E., Amouroux J.-M., Jabouin C., Gigou A. *et al.*, 2019. The arrival of the American blue crab, *Callinectes sapidus* Rathbun, 1896 (Decapoda: Brachyura: Portunidae), in the Gulf of Lion (Mediterranean Sea). *BioInvasions Records*, 8 (4), 876–881. https://doi.org/10.3391/bir.2019.8.4.16

Lassus P., Chomérat N., Hess P., Nézan E., 2016. Toxic and harmful microalgae of the World Ocean. Denmark, International Society for the Study of Harmful Algae/Intergovernmental Oceanographic Commission of Unesco. IOC Manuals and guides, 68 (bilingual english/french), 523 p.

Le Gouvello R., Simard F., 2024. *Vers une économie bleue régénérative. Une cartographie de l'économie bleue*, IUCN, Gland, Switzerland, 58 p.

Lehtiniemi M., Ojaveer H., David M., Galil B., Gollasch S. *et al.*, 2015. Dose of thruth: monitoring marine non-indigenous species to serve legislative requirements. *Marine Policy*, 54, 26–35. https://doi.org/10.1016/j.marpol.2014.12.015

Leray M., Knowlton N., Machida R.J., 2022. MIDORI2: a collection of quality controlled, preformatted, and regularly updated reference databases for taxonomic assignment of eukaryotic mitochondrial sequences. *Environmental DNA*, 4 (4), 894–907. https://doi.org/10.1002/edn3.303

Leroy B., Kramer A.M., Vaissière A., Kourantidou M., Courchamp F. *et al.*, 2022. Analysing economic costs of invasive alien species with the INVACOST R package. *Methods in Ecology and Evolution*, 13 (9), 1930–1937. https://doi.org/10.1111/2041-210x.13929

Lindroth C.H., 1957. *The Faunal Connections between Europe and North America*, Wiley, New York, 345 p.

Llinares S., Egasse B., Dana K., 2018. *De l'estran à la digue : histoire des aménagements portuaires et littoraux XVIe -XXe siècle*, Presses universitaires de Rennes, 416 p. https://doi.org/10.4000/books.pur.173767

Maes J., Teller A., Condé S. *et al.*, 2020. *Mapping and Assessment of Ecosystems and their Services. An EU Wide Ecosystem Assessment in Support of the EU Biodiversity Strategy*, Publications Office, 452 p. + appendices. https://doi.org/10.2760/757183

Maes C., 2022. Dispersion des plastiques marins flottants à la surface des océans. *La Météorologie*, 119, 53–61.

Magliozzi C., Gervasini E., Cardoso A.C., 2024. *Informing Spatiotemporal Trends of Invasive Alien Species of Union Concern with Biological Knowledge*, Publications Office of the European Union, Luxembourg. https://doi.org/10.2760/298316

Mainka S.A., Howard G.W., 2010. Climate change and invasive species: double jeopardy. *Integrative Zoology*, 5, 102–111. https://doi.org/10.1111/j.1749-4877.2010.00193.x

Maric M., Troch M.D., Occhipinti-Ambrogi A., Olenin S., 2016. Trophic interactions between indigenous and non-indigenous species in Lampedusa Island, Mediterranean Sea. *Marine Environmental Research*, 120, 182–190. https://dx.doi.org/10.1016/j.marenvres.2016.08.005

Markert A., Esser W., Frank D., Whermann A., Exo K.M., 2013. Habitat change by the formation of alien *Crassostrea* reefs in the Wadden Sea and its role as feeding sites for waterbirds. *Estuarine and Coastal Shelf Science*, 131, 41–51. https://doi.org/10.1016/j.ecss.2013.08.003

Martel C., Viard F., Bourguet D., Garcia-Meunier P., 2004a. Invasion by the marine gastropod *Ocinebrellus inornatus* in France. I. Scenario for the source of introduction. *Journal of Experimental Marine Biology and Ecology*, 305, 155–170.

Martel C., Viard F., Bourguet D., Garcia-Meunier P., 2004b. Invasion by the marine gastropod *Ocinebrellus inornatus* in France. II. Expansion along the Atlantic Coast. *Marine Ecology Progress Series*, 273, 163–172.

Martínez-García M.F., Ruesink J.L., Grijalva-Chon J.M., Lodeiros C., Arreola-Lizarraga J.A. *et al.*, 2021. Socioecological factors related to aquaculture introductions and production of Pacific oysters (*Crassostrea gigas*) worldwide. *Reviews in Aquaculture*, 14 (2), 613–629. https://doi.org/10.1111/raq.12615

Massé C. *et al.*, 2023. An overview of marine non-indigenous species found in three contrasting biogeographic metropolitan French regions: insights on distribution, origins and pathways of introduction. *Diversity*, 15, 161. https://doi.org/10.3390/d15020161

Mathieu J., Reynolds J.W., Fragoso C., Hadly E., 2024. Multiple invasion routes have led to the pervasive introduction of earthworms in North America. *Nature Ecology and* Evolution. https://doi.org/10.1038/s41559-023-02310-7

Meinard Y., Dereniowska M., Glatron S., Maris V., Phelippot V. *et al.*, 2022. A heuristic for innovative invasive species management actions and strategies. *Ecology and Society*, 27 (4), 24. https://doi.org/10.5751/ES-13615-270424

Meinesz A. *et al.*, 2001. The introduced green alga *Caulerpa taxifolia* continues to spread in the Mediterranean. *Biological Invasions*, 3, 201–210.

Meyer J.C., 2023. Le miconia, "cancer vert" des forêts du Pacifique. *In: 50 ans de recherche pour le développement en Polynésie* (P. Lacombe *et al.*, eds), IRD Open Éditions, 58–67. 10.4000/books.irdeditions.44270

Mghili B., De-la-Torre G.E., Aksissou M., 2023. Assessing the potential for the introduction and spread of alien species with marine litter. *Marine Pollution Bulletin*, 191, 114913.

MEA, 2005. *Ecosystems and Human Well-being: Synthesis*, Millennium Ecosystem Assessment, Island Press, Washington. https://www.millenniumassessment.org/en/index.html

Mineur F., Le Roux A., Maggs C.A., Verlaque M., 2014. Positive feedback loop between introductions of non-native marine species and cultivation of oysters in Europe. *Conservation Biology*, 28 (6), 1667–1676. https://doi.org/10.1111/cobi.12363

Nasco Council, 2022a. Distribution and abundance of pink salmon *Oncorhynchus gorbuscha* across the North Atlantic. ICES Advice on fishing opportunities, catch, and effort, North Atlantic Ecoregions, CNL (22) 64, 16 p.

Nasco Council, 2022b. Statement of the Council regarding pink salmon, *Oncorhynchus gorbuscha* , in the NASCO Convention Area. CNL (22) 47, 2 p.

Nasco Council, 2023. Draft Terms of Reference for the NASCO Working Group on Pink Salmon *Oncorhynchus gorbuscha*. CNL (23) 26, 3 p.

Normant-Saremba M., 2024. The Chinese mitten crab *Eriocheir sinensis* is spreading microsporidian parasite *Hepatospora eriocheir* across Europe. Could it be beneficial? *In: ICAIS Conference*, May 2024.

Novi L., Braccol A., Falasca F., 2021. Uncovering marine connectivity through sea surface temperature. *Scientific Reports*, 11, 8839. https://doi.org/10.1038/s41598-021-87711-z

Nunes F., Bouchoucha M., Carlier A., Curd A., Droual G. *et al.*, 2024. Integrative taxonomy as an essential part of detecting and monitoring non-indigenous species with DNA-based methods: examples from the Northeastern Atlantic and Mediterranean. *In: ICAIS Conference, May 2024, Halifax.*

OFB, 2024. https://www.ofb.gouv.fr/les-especes-exotiques-envahissantes

Ojaveer H., Galil B.S., Campbell M.L., Carlton J.T., Canning-Clode J. *et al.*, 2015. Classification of non-indigenous species based on their impacts: considerations for application in marine management. *PLoS Biology*, 13 (4), e1002130. https://doi.org/10.1371/journal.pbio.1002130

Ojaveer H., Gallil B., Carlton J.T., Allewayn H., Goulletquer P. *et al.*, 2018. Historical baselines in marine bioinvasions: implications for policy and management. *PLoS ONE*, 13 (8), e0202383. https://doi.org/10.1371/journal.pone.0202383

Olden J., 2024. Public bounty programmes to control aquatic invasive species. *In: ICAIS Conference*, May 2024.

Olenin *et al.*, 2009. Good Environmental Status (GES) descriptor: non-indigenous species introduced by human activities are at levels that do not adversely affect ecosystems. EU Commission, Joint Research Center.

Olenin *et al.*, 2011. Recommendations on methods for the detection and control of biological pollution in marine coastal waters. *Marine Pollution Bulletin*, 62, 2598–2604. http://dx.doi.org/10.1016/j.marpolbul.2011.08.011

Olenin *et al.*, 2014. Making non-indigenous species information systems practical for management and useful for research: an aquatic perspective. *Biological Conservation*, 173, 98–107.

Outinen O., Bailey S.A., Broeg K. *et al.*, 2021. Exceptions and exemptions under the ballast water management convention: sustainable alternatives for ballast water management? *Journal of Environmental Management*, 293112823. https://doi.org/10.1016/j.jenvman.2021.112823

Pante E., Schoelinck C., Puillandre N., 2015. From integrative taxonomy to species description: one step beyond. *Systematic Biology*, 64 (1), 152–160. https://www.doi.org/10.1093/sysbio/syu083

Pastor A., Catalan I.A., Terrados J., Mourre B., Ospina-Alvarez A., 2023. Connectivity-based approach to guide conservation and restoration of seagrass *Posidonia oceanica* in the NW Mediterranean. *Biological Conservation*, 110248. https://doi.org/10.1016/j.biocon.2023.110248

Pépin J.F., 2023. Le projet de recherche RAPSODI. https://littoral.ifremer.fr/Laboratoires-Environnement-Ressources/LER-Pertuis-Charentais-La-Tremblade/Projets/RAPSODI-2022-2024

Pereira L., Davies K.K., den Belder E., Ferrier S. et al., 2020. Developing multiscale and integrative nature-people scenarios using the Nature Futures Framework. *People and Nature*, 2, 1172–1195. https://doi.org/10.1002/pan3.10146

Pérez-Granados C., Lenzner B., Golivets M., Sauf W.C. et al., 2023. European scenarios for future biological invasions. *People and Nature*, 6, 245–259. https://doi.org/10.1002/pan3.10567

Perriman B.M., Bentzen P., Wringe B.F., Duffy S., Islam S.S. et al., 2022. Morphological consequences of hybridization between farm and wild Atlantic salmon *Salmo salar* under both wild and experimental conditions. *Aquaculture Environment Interactions*, 14, 85–96. https://doi.org/10.3354/aei00429

Piazzi L., Meinesz A., Verlaque M., Akali B., Antolic B. et al., 2005. Invasion of *Caulerpa racemosa* var. *cylindracea* (Caulerpales, Chlorophyta) in the Mediterranean Sea: an assessment of the spread. *Cryptogamie, Algologie*, 26, 189–202.

Picciotto M., Bertuccio C., Glacobbe S., Spano N., 2016. *Caulerpa taxifolia* var. *distichophylla*: a further stepping stone in the western Mediterranean. *Marine Biodiversity Records*, 9, 73. https://doi.org/10.1186/s41200-016-0038-1

Pigeot J.T., Miramand P., Garcia-Meunier P., Guyot T.T., Seguignes M., 2000. Presence of a new predator of the eastern oyster *Ocinebrellus inornatus* (Récluz, 1851), in the Marennes-Oléron shellfish basin. *Comptes rendus de l'Académie des sciences. Série III, Sciences de la vie*, 323 (8), 697–703.

Pisanu B., Massé C., Thévenot J., Bachelet G., Bierne N. et al., 2020. Proposition d'espèces non-indigènes pour les façades maritimes du territoire métropolitain à soumettre à réglementation. UMS Patrimoine naturel, 18 p.

Pitois S., Shiganova T., 2015. Report of the Joint ICES/ICES Workshop on *Mnemiopsis*. Science-A Coruna, Spain, 80 p.

Pombo A., Baptista T., Granada L., Ferreira S.M.F., Gonçalves S.C. et al., 2020. Insight into aquaculture's potential of marine annelid worms and ecological concerns: a review. *Reviews in Aquaculture*, 12, 107–121. https://doi.org/10.1111/raq.12307

Ratnasingham S., Hebert P.D.N., 2007. The Barcode of Life Data System. *Molecular Ecology Notes*, 7 (3), 355–364. https://doi.org/10.1111%2Fj.1471-8286.2007.01678.x

Rémy E., Beck C., 2008. Allochtone, autochtone, invasif : catégorisations animales et perception d'autrui. *Politix*, 2 (82), 193–209.

Ricciardi A., Iacarella J.C., Aldridge D.C., Blackburn T.M., Carlton J.T. *et al.*, 2020. Four priority areas to advance invasion science in the face of rapid environmental change. *Environmental Reviews*, 29 (2) 119–141. https://doi.org/10.1139/er-2020-0088

Riley S., Molina V., First M., 2024. Optical approaches to quantify biofouling. ICAIS, Halifax, Abstract 240.

Rishan S.T., Kline R.J., Rahman M.dS., 2023. Applications of environmental DNA (eDNA) to detect subterranean and aquatic invasive species: a critical review on the challenges and limitations of eDNA metabarcoding. *Environmental Advances*, 12100370. https://doi.org/10.1016/j.envadv.2023.100370

Ruitton S., Blanfuné A., Boudouresque C.-F., Guillemain D., Michotey V. *et al.*, 2021. Rapid spread of the invasive Brown Alga *Rugulopteryx okamurae* in a National Park in Provence (France, Mediterranean Sea). *Water*, 13, 2306. https://doi.org/10.3390/w13162306

Sa E., Fidalgo e Costa P., Cancela de Fonseca L., Alves A.S., Castro N. *et al.*, 2017. Trade of live bait in Portugal and risks of introduction of non-indigenous species associated to importation. *Ocean and Coastal Management*, 146, 121–128.

Sauriau P.G., Dartois M., Becquet V., Aubert F., Huet V. *et al.*, 2021. Multiple genetic marker analysis challenges the introduction history of *Ulva australis* (Ulvales, Chlorophyta) on French coasts. *European Journal of Phycology*, 13 p. https://doi.org/10.1080/09670262.2021.1876249

Schaber M., Haslob H., Huwer B. *et al.*, 2011. The invasive ctenophore *Mnemiopsis leidyi* in the central Baltic Sea: seasonal phenology and hydrographic influence on spatio-temporal distribution patterns. *Journal of Plankton Research*, 33 (7), 1053–1065. https://doi.org/10.1093/plankt/fbq167

Sellier M., Poitevin P., Goraguer H., Faure J.M., Goulletquer P., 2016. Suivi des espèces envahissantes marines à Saint-Pierre-et-Miquelon. Année 2014. Convention n° 254 du 19/06/2014 modifiant la n° 210 du 25/05/2014. https://archimer.ifremer.fr/doc/00312/42303/

Shakspeare A., Cameron T.C., Steinke M., 2024. Restrictions on UK aquaculture of Pacific oyster (*Magallana gigas*) will not prevent naturalised spread but suppress ecological and economic benefits to coastal communities. A critical review. EcoEvoRxiv. https://doi.org/10.32942/X2PP60

Shea K., Chesson P., 2002. Community ecology theory as a framework for biological invasions. *Trends in Ecology and Evolution*, 17 (4), 170–176. https://doi.org/10.1016/S0169-5347(02)02495-3

Simberloff D., Martin J.L., Genovesi P., Maris V., Wardle D. *et al.*, 2013. Impacts of biological invasions: what's what and the way forward. *Trends in Ecology and Evolution*, 28 (1), 58–66. https://doi.org/10.1016/j.tree.2012.07.013

Simberloff D., Von Holle, 1999. Positive interactions of non-indigenous species: invasional meltdown? *Biological Invasions*, 1 (1), 21–32. https://doi.org/10.1023/A:1010086329619

Simon-Bouhet B., Garcia-Meunier P., Viard F., 2006. Multiple introductions promote range expansion of the mollusc *Cyclope neritea* (Nassaridae) in France: evidence from mitochondrial sequence data. *Molecular Ecology*, 15, 1699–1711.

Smanski M. *et al.*, 2024. Progress towards genetic biocontrol of invasive carp. *In: ICAIS Conference*, May 2024.

SMEL, 2016. Synergy, sea and coast. Projet SNOTRA 2- https://www.smel.fr/download-category/rapport-final/

Smith K.G., Nunes A.L., Aegerter J., Baker S.E., Di Silvestre I. *et al.*, 2022. A manual for the management of vertebrate invasive alien species of Union concern, incorporating animal welfare, 1rst edition. Technical report prepared for the European Commission within the framework of the contract no 07.027746/2019/812504/SER/ENV.D.2.

SNB3, 2024. National Biodiversity Strategy 3. https://www.ecologie.gouv.fr/politiques-publiques/strategie-nationale-biodiversite-2030

Soto I., Palzani P., Carneiro L. *et al.*, 2024. Taming the terminological tempest in invasion science. *Biological Reviews*, 99 (4), 1357–1390. https://doi.org/10.1111/brv.13071

Sparfel L., Fichaut B., Suanez S., 2005. Progression de la Spartine (*Spartina alterniflora* Loisel) en rade de Brest (Finistère) entre 1952 et 2004 : de la mesure à la réponse gestionnaire. *Norois*, 196 (3), 109–123. https://doi.org/10.4000/norois.438

Strayer D.L., Eviner V.T., Jeschke J.M., Pace M.L., 2006. Understanding the long-term effects of species invasions. *Trends in Ecology and Evolution*, 21 (11), 645–651.

Tan N., Miller J.A., Chapman J.W., Pleus A.E., Calvanese T. *et al.*, 2018. The Western Pacific barred knifejaw *Oplegnathus fasciatus* arriving with tsunami debris on the Pacific coast of North America. *Aquatic Invasions*, 13 (1), 179–186.

Therriault T., Nelson J., Carlton J., Liggan L., Otani M. *et al.*, 2018. The invasion risk of species associated with Japanese tsunami marine debris in Pacific North America and Hawai. *Marine Pollution Bulletin*, 132, 82–89. https://doi.org/10.1016/j.marpolbul.2017.12.063

Thomas Y., Pouvreau S., Alunno-Bruscia M., Barillé L., Gohin F. *et al.*, 2016. Global change and climate-driven invasion of the Pacific oyster along European coasts: a bioenergeticcs modelling approach. *Journal of Biogeography*, 43, 568–579. https://doi.org/10.1111/jbi.12665

Tournadre J., 2014. Anthropogenic pressure on the open ocean: the growth of ship traffic revealed by altimeter data analysis. *Geophysical Research Letters*, 41, 7924–7932. https://doi.org/10.1002/2014GL061786

Tsirintanis K., Azzuro E., Crocetta F., Dimiza M., Froglia C. *et al.*, 2022. Bioinvasion impacts on biodiversity, ecosystem services and human health in the Mediterranean Sea. *Aquatic Invasions*, 17 (3), 308–352.

Tsirintanis K., Sini M., Ragkousis M., Zenetos A., Katsanevakis S., 2023. Cumulative negative impacts of invasive alien species on marine ecosystems of the Aegean Sea. *Biology*, 12 (7), 933. https://doi.org/10.3390/biology12070933

Turbelin A.J., Cuthbert R.N., Essl F., Haubrock P.J., Ricciardi A. *et al*., 2023. Biological invasions are as costly as natural hazards. *Perspectives in Ecology and Conservation*, 21 (2), 143–150. https://doi.org/10.1016/j.pecon.2023.03.002

UNEP, 2021. *Comprehensive Assessment on Marine Litter and Plastic Pollution Confirms Need for Urgent Global Action*, Nairobi, 148 p.

US EPA, 2008. Effects of climate change for aquatic invasive species and implication for management and research. Environmental Protection Agency. National Service Center for Environmental Publications (NSCEP), https://assessments.epa.gov/risk/document/&deid%3D188305

Vitale D., 2017. The Suez Canal as a revolving door for marine species: a reply to Galil *et al.* (2016). *Aquatic Invasions*, 12 (1), 1–4.

Vilizzi L. *et al*., 2021. A global-scale screening of non-native aquatic organisms to identify potentially invasive species under current and future climate conditions. *Journal Science of the Total Environment*, 788, 147868. https://doi.org/10.1016/j.scitotenv.2021.147868; Application download site: https://siren.fort.usgs.gov/static-page/invasiveness-screening-kit-decision-support-tools-for-the-identification-and-management-of-invasive-non-native-species

Vilizzi L. *et al*., 2025. To be, or not to be, a non-native species in non-English languages: gauging terminological consensus amongst invasion biologists. *Management of Biological Invasions*, 16(1):15–31.

Voisin M., Daguin C., Engel C., Grulois D., Javanaud C. *et al*., 2007. Processus et dynamique d'installation des espèces introduites en milieu marin / une illustration avec l'algue brune asiatique *Undaria pinnatifida*. *Journal de la Société de biologie*, 201 (3), 259–266.

Warren C.R., 2021. Beyond 'Native vs. Alien': critiques of the native/alien paradigm in the Anthropocene, and their implications. *Ethics, Policy and Environment*, 26 (2), 287–317. https://doi.org/10.1080/21550085.2021.1961200

Waser A., Deuzeman S., Kangeri A.K., van Winden E., Postma J. *et al*., 2016. Impact on bird fauna of a non-native oyster expanding into the blue mussel beds in the Dutch Wadden Sea. *Biological Conservation*, 202, 39–49. https://doi.org/10.1016/j.biocon.2016.08.007

Willan R.C., Russell B.C., Murfet N.B., Moore K.L., McEnnulty F.R. *et al*., 2000. Outbreak of *Mytilopsis sallei* (Récluz, 1849) in Australia. *Molluscan Research*, 20 (2), 25–30.

Wolff W.J., Reise K., 2002. Oyster imports as a vector for the introduction of alien species into Northern and Western European coastal waters. *In: Invasive Aquatic Species of Europe. Distribution, Impacts and Management* (E. Leppäkoski *et al*., eds), Springer, 193–205. https://link.springer.com/chapter/10.1007/978-94-015-9956-6_21

Wonham M.J., Lewis M.A., 2009. Modeling marine invasions: current and future approaches. *In: Biological Invasions in Marine Ecosystems* (G. Rilov, J.A. Crooks, eds). *Ecological Studies*, 204, Springer, Berlin, Heidelberg. https://doi.org/10.1007/978-3-540-79236-9_4

Wotton D.M., O'Brien C., Stuart M.D., Fergus D.J., 2004. Eradication success down under: treatment of a sunken trawler to kill the invasive seaweed *Undaria pinnatifida*. *Marine Pollution Bulletin*, 49 (9–10), 844–849. https://doi.org/10.1016/j.marpolbul.2004.05.001

Zeni R.D., Essl F., Garcia-Berthou E., McDernmott S.M., 2021. The economic costs of biological invasions around the world. *Neobiota*, 67, 1–9. https://doi.org/10.3897/neobiota.67.69971

www.ingramcontent.com/pod-product-compliance
Ingram Content Group UK Ltd.
Pitfield, Milton Keynes, MK11 3LW, UK
UKHW022139110226
467951UK00001B/1